Springer Series in

MATERIALS SCIENCE 131

Springer Series in

MATERIALS SCIENCE

The Springer Series in Materials Science covers the complete spectrum of materials physics, including fundamental principles, physical properties, materials theory and design. Recognizing the increasing importance of materials science in future device technologies, the book titles in this series reflect the state-of-the-art in understanding and controlling the structure and properties of all important classes of materials.

Please view available titles in *Springer Series in Materials Science* on series homepage http://www.springer.com/series/856

Kazuko Motizuki • Hideaki Ido
Tadaei Itoh • Masato Morifuji

Electronic Structure and Magnetism of 3d-Transition Metal Pnictides

With 118 Figures

Springer

Prof. Kazuko Motizuki
1-10-16 Nigawa-takadai
Takaraduka
665-0062, Japan

Prof. Hideaki Ido
Prof. Tadaei Itoh
Tohoku Gakuin University
Dept. Electronic Engineering
1-13-1 Chuo
Tagajo, Miyagi
985-0873, Japan

Prof. Masato Morifuji
Osaka University
Dept. Quantum Electronic
Device Engineering
2-1 Yamada-oka
Suita, Osaka
565-0871, Japan

Series Editors

Springer Series in Materials Science ISSN 0933-033X
ISBN 978-3-642-03419-0 e-ISBN 978-3-642-03420-6
DOI 10.1007/978-3-642-03420-6
Springer Heidelberg Dordrecht London New York

Library of Congress Control Number: 2009933995

The original Japanese edition was published by Daigaku Kyouiku Syuppan, Okayama, Japan, 2007
Original title: *Kinzokukan Kagoubutsu no Denshikouzou to Jisei*

Cover design: SPi Publisher Services

Printed on acid-free paper

Springer is part of Springer Science+Business Media (www.springer.com)

Preface

This book describes in 2 parts experimental data with simple explanations (Part I) and itinerant electron theories (Part II) about magnetism and its related properties of $3d$-intermetallic compounds. Unlike $3d$-metal alloys and oxides, theoretical as well as experimental studies on $3d$-intermetallic compounds such as $3d$-pnictides and chalcogenides, on which we focus in this book, seem unfortunately delayed. The objective of this book is to motivate active studies in this field in the future.

We discuss in detail magnetic and related properties of the $3d$-transition-metal pnictides and chalcogenides, which include the intermetallic compounds expressed as MX and M_2X, and their mixed compounds $M_{1-x}M'_xX$, $MX_{1-y}X'_y$ and $M_{2-x}M'_xX$, where M (M′) is a $3d$ element and X (X′) a pnicogen (P, As, Sb, and Bi) or a chalcogen (S, Se, and Te). Most of the MX-type compounds crystallize either in the hexagonal NiAs-type structure or in the orthorhombic MnP-type structure which is regarded as a distorted NiAs-type structure. Crystallographic phase transition between the NiAs- and the MnP-types occurs in some of MX-type compounds when the temperature changes. The M_2X-type compounds crystallize in the tetragonal Cu_2Sb-type structure. As described in detail in this book, many of the compounds mentioned above exhibit very interesting magnetic and crystallographic phase transitions caused by various means such as change of temperature, applications of external magnetic field or pressure, and change of the composition x in the case of mixed compounds. This variety of phase transitions seems to have motivated many researchers to investigate the group of compounds mentioned above. We, one experimentalist and three theorists, had been collaborating for a long time to clarify the mechanism of phase transitions as well as the peculiar magnetic properties appearing in the group of compounds mentioned above. Considerable progress in understanding the properties of the compounds mentioned above has been achieved; however, there still remain various phenomena not clarified yet. Besides the fundamental problems, those of potential applications are also considered for the phenomena observed in some of the compounds mentioned above.

In Part I the crystallographic and magnetic properties are described together with simple phenomenological analyses the consideration of applications is given in a few sections. In Chap. 1 we show the method to prepare a polycrystalline sample, crystal structures and the phase diagram of a typical MX-type compound. In Chap. 2 we survey basic magnetic data such as magnetic transition temperatures, magnetic moments and magnetic orders, etc. for the compounds of MX- and M_2X-types. Chapter 3, the main part in Part I, presents detailed descriptions of magnetic and related properties for respective compounds. In Sects. 3.1–3.6 properties of the compounds containing Mn or Cr, which are the most interesting from both the fundamental and application view points, are described in detail, together with phenomenological theory to explain magnetic properties, and application of MnAs as a magnetic refrigerant is also included in these sections. Properties of the M_2X-type compounds are shown in Sect. 3.7. Many pages of this section are used for the Cr-modified Mn_2Sb.

Part II, where theoretical results are shown, is divided further in to two chapters. In Chap. 4 theoretical results of the NiAs-(MnP)-type compounds are shown. The magnetism of materials arises from $3d$ or $4f$ electrons. There are two different models to describe the magnetic electrons: the localized model and the itinerant electron model. Theories shown in Part II are based on the itinerant electron picture. In Sects. 4.1 and 4.2 we show and discuss in detail electronic band structures calculated by the first principle method for paramagnetic, ferromagnetic or antiferromagnetic states of the MX-type compounds in the $3d$-pnictides and chalcogenides. In Sects. 4.3–4.7 we theoretically investigate various experimental data on the basis of the calculated electronic band structures. In Sect. 4.3 we clarify origin of the anomalous behavior of paramagnetic susceptibility and thermal expansion of MnAs by taking the spin fluctuation into account. Correspondence between Fermi surfaces and de Haas-van Alphen oscillations in NiAs and NiSb is discussed in Sect. 4.4. Pressure induced magnetic transition in CrTe is discussed in Sect. 4.5. In Sect. 4.6 we present a theory to search for the most stable magnetic ordering in FeAs, where instability of the paramagnetic state is investigated from the wavevector-dependent susceptibility tensor. In the final Sect. 4.7 structural transformation from the NiAs-type to the MnP-type structures is discussed. On the basis of the calculated band structures, we investigate the instability of the NiAs-type structure against MnP-type lattice distortion by taking the electron-lattice interaction into account.

In Chap. 5 we show the theoretical results of the Cu_2Sb-type compounds. Crystal structure and magnetic ordering are shown in Sect. 5.1. In Sect. 5.2 band structures of the Cu_2Sb-type compounds calculated by the first principle method are shown, followed by a comparison between experimental and theoretical results in magnetic ordering and optical spectra.

This book is the result of a longlasting close collaboration between experimentalists and theorists, which is also the tradition of the study of magnetism in Japan. It would be our great pleasure if this book proves helpful to graduate students, researchers and engineers in the field of science and technology of magnetism.

Miyagi, Osaka
October 2009

Kazuko Motizuki
Hideaki Ido
Tadaei Itoh
Masato Morifuji

Contents

Part I Experimental

Part I

Experimental

1

Basic Properties of 3*d*-Pnictides

1.1 Main Compounds of 3*d*-Pnictide

Among various intermetallic compounds between 3*d* transition metal M and pnicogen X (X = P, As, Sb, or Bi), compounds expressed as MX and M_2X are of special interest due to their rich variety of magnetic properties. The MX compounds crystallize basically in the hexagonal NiAs-type ($B8_1$) structure, however, some of them show distortion into the orthorhombic MnP-type (B31) structure. The M_2X compounds take the tetragonal Cu_2Sb-type structure. The MX and M_2X compounds are listed in Table 1.1.

In this table, the MX compounds denoted by boldface letters take the NiAs- or MnP-type structure. The M_2X compounds written in boldface take the Cu_2Sb-type structure. It should be pointed out that the ternary compounds MnAlGe and MnGaGe also have the Cu_2Sb-type structure. Some compounds take nonstoichiometric composition such as $M_{1+x}X$ ($x \geq 0$) although they are written as MX in Table 1.1 [1]. For example, MnSb actually forms a compound $Mn_{1+x}Sb$ with $0 < x < 0.2$. Some other NiAs-type compounds also have similar tendency [1]. As the excessive Mn atoms, which occupy interstitial positions, have a strong effect on magnetic properties, a special attention should be paid for preparation of stoichiometric MnSb compound [2]. For the NiAs-type structure, we can regard that M atoms with smaller size than that of X atoms occupy interstitial positions of the hexagonal close-packed structure formed first by the X atoms. Therefore, this may permit existence of excessive M atoms. Among the compounds shown in Table 1.1, only Cr-, Mn-, and Fe- compounds take magnetic ordering.

1.2 Crystal Growth by Sintering Method

Crystal growth is the starting point of experimental studies of solid state physics. Although there has been rapid progress in equipment and technique, we show here a conventional method to prepare a polycrystalline sample. This method is called

Table 1.1. Compounds between 3*d* metal and pnicogen

	P	As	Sb	Bi
Ti	TiP	TiAs	**TiSb**	–
V	**VP**	**VAs**	**VSb**	–
Cr	**CrP**	**CrAs, Cr_2As**	**CrSb**	–
Mn	**MnP**, Mn_2P	**MnAs, Mn_2As**	**MnSb, Mn_2Sb**	**MnBi**
Fe	**FeP**, Fe_2P	**FeAs, Fe_2As**	**FeSb**	–
Co	**CoP**, Co_2P	**CoAs**	**CoSb**	–
Ni	NiP	**NiAs**	**NiSb**	**NiBi**
Cu	–	–	**Cu_2Sb**	–

The compounds written in boldface have the NiAs-, MnP-, or Cu_2Sb-type structure

a sintering method, in which mixture of powders of constituent elements settled in an evacuated silica tube is sintered in a furnace. This method is applicable for most intermetallic compounds.

Usually particle size of metal powder is about 100 μm, but larger or smaller particle sizes can be acceptable. Mn can be ground easily in a mortar of agate; however, other 3*d* metals are very difficult to pulverize. It is recommended to obtain the powdered metal from specialty store. Pnicogen is ground easily, but special care is necessary for phosphorus and arsenic. As for phosphorus, we use red phosphorus (not white phosphorus), which may explode while grinding. One can purchase powder of red phosphorus. Since As is toxic (As_2O_3 is deadly poisonous), one must wear mask not to inhale the fine particles and be careful so that particles do not scatter. Tidying up after experiments is also necessary. Purity of all ingredient should be more than 99.9%.

We show procedure of sintering for the case of MnAs. First, we wash electrolytic manganese in dilute nitric acid and then in water. Next we grind it in a mortar of agate. High purity bulk of As taken out of a vacuum glass container is crushed into powder. Powders of Mn and As are weighted in desired proportion and mixed in a mortar carefully so that the powder does not scatter. These processes should be finished as quickly as possible to avoid oxidation of the powdered metals. Then, the mixture of Mn and As (total weight of 5–8 g) is sealed in an evacuated quartz tube. An example of shape of the quartz tube is shown in Fig. 1.1.

The neck of the quartz tube should be carefully prepared. If this part is too thin, keeping the tube in high vacuum may fail, and ingredient powder cannot go inside through the neck with too small internal diameter.

The mixture sealed in the evacuated quartz tube is heated in an electric furnace. Since As sublimates at 613°C and has large vapor pressure at high temperature, furnace temperature must be raised slowly spending a few days up to 500°C in order to avoid explosion. Then the temperature is kept constant for another few days so that rough reaction of the mixture finishes. In the final stage, the temperature is raised up to about 800°C (you can heat up rapidly this time) and kept for a week. We can take out the reacted product after cooling down. As the sintered crystal thus prepared

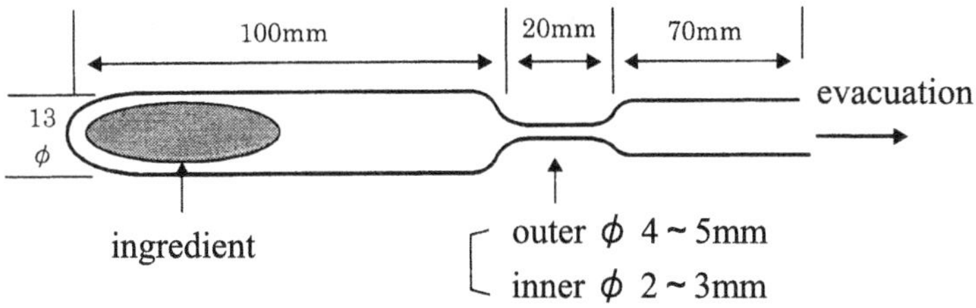

Fig. 1.1. A sample tube of quartz. After evacuating up to about 10^{-5} Torr, the neck of the quartz tube is fused and cut by a gas burner. The dimensions shown in this figure are typical ones

may be inhomogeneous, the sintering at 800°C for a week is desirable to be repeated. Namely, the reaction product is pulverized and sealed again in an evacuated quartz tube. One can heat the sample rapidly this time. Longer the period of sintering, better the quality of the crystal.

As phosphorus also has risk of explosion, the procedure mentioned earlier must be taken for compounds containing P. Other compounds also can be prepared in the similar way. The diagram of binary alloys [3] will be helpful for determining the sintering temperature.

1.3 Crystal Structure of NiAs-, MnP-, and Cu_2Sb-Type Structure

1.3.1 NiAs-Type Structure

As the MnP-type structure is regarded as distorted NiAs-type structure with small shifts of atomic positions, these two structures are similar to each other. We explain below these crystal structures connecting them with each other.

Figure 1.2 shows the hexagonal NiAs-type structure. Pnicogen atoms form a lattice like the hexagonal close-packed structure, and 3*d* metal atoms a simple hexagonal lattice. Table 1.2 shows lattice constants of the MnX compounds at room temperature.

As seen in Table 1.2, with change of pnicogen from Bi to P, the lattice constants decrease. In particular, decrease of the *a*-axis is remarkable. It is noted that *a*-axis of MnP is 73% of that of MnBi, corresponding to the fact that radius of P atom is 75% of that of Bi. Hence we see that the *a*-axis is almost proportional to radius of the pnicogen atom. Such tendency is also found in the compounds CrP, CrAs, and CrSb.

1.3.2 MnP-Type Structure

The MnP-type structure is obtained by introducing an orthorhombic distortion as well as slight shifts of atom positions for the hexagonal NiAs-type structure. As distortion and shifts are small, these two structures are basically similar. However, even the slight difference affects magnetic properties through change of electronic structure.

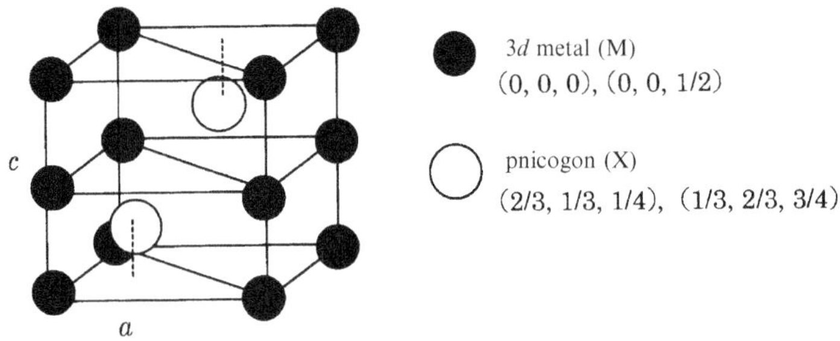

Fig. 1.2. The NiAs-type structure (B81)

Table 1.2. Lattice constants *a* and *c* (room temperature) of MnX [4] and radius of the X atom

	(MnP)	MnAs	MnSb	MnBi
a (Å)	3.17	3.68	4.15	4.34
c (Å)	5.26	5.72	5.78	5.98
Atom radius of X (Å)	1.28	1.39	1.59	1.70

Note that MnP takes the orthorhombic MnP-type structure (*a* and *c* are lattice constants corresponding to those of the hexagonal NiAs-type structure)

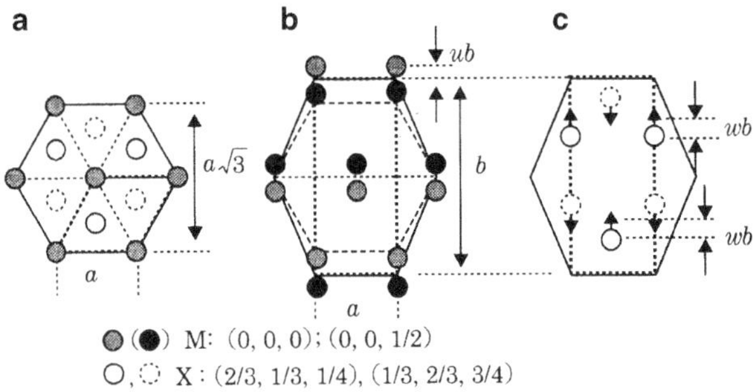

Fig. 1.3. Relation between the NiAs-type and the MnP-type structure projected onto the *c*-plane. (**a**) The NiAs-type structure (The *thick lines* denote the unit cell. See also Fig. 1.2.) (**b**) Distortion into the orthorhombic lattice and displacement of M atoms along the *b*-axis are shown. The displacement is written as *ub* using a displacement parameter *u*. (**c**) Displacement of X atoms in the *b*-direction. The displacement is written as *wb* (*w*: a displacement parameter) whose origin is 1/3 or 1/6 of the *b*-axis. Note that the displacements and distortion are enlarged in order for the illustration to be legible. See also Fig. 4.38

We explain the difference between the MnP-type and the NiAs-type structure. Figure 1.3 shows atomic positions projected onto the *c*-plane for the NiAs-type and MnP-type structures, where displacements of atoms in the *c*-plane are also illustrated.

In Fig. 1.3, crystallographic parameters u and w indicate atomic displacement along the b-axis. The parameter w is one order smaller than u. Distortion from the NiAs-type to MnP-type structure is expressed with a parameter δ, which denotes deviation of the b-axis from $\sqrt{3}a$-axis :

$$\delta = (b - \sqrt{3}a)/\sqrt{3}a.$$

Atoms also shift along the c-axis as shown in Fig. 1.4.

In summary, the MnP-type structure is characterized by the five crystallographic parameters: the displacement parameters of metal (u and x) and anion (v and w) and the distortion parameter (δ). The NiAs-type structure is thus expressed by setting $u = v = x = w = \delta = 0$. The crystallographic parameters are listed in Table 1.3 for MnP and CrAs, which take the MnP-type structure.

Other MnP-type compounds have similar values of the parameters. Since w and x are generally smaller than u and v, only u, v, and δ can be determined in

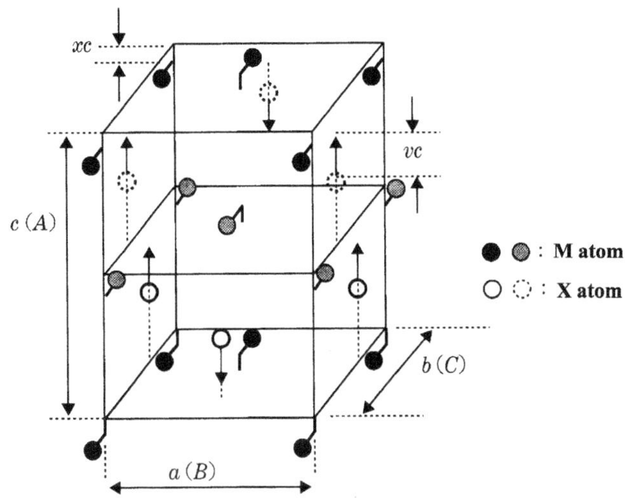

Fig. 1.4. Unit cell of the MnP-type structure. Displacements of M atoms and X atoms in the $c(A)$-direction are given by xc and vc, respectively. The parameter x is smaller than v by one order. In the expression of lattice parameters of the MnP-type structure, (A, B, C) is often used in addition to (a, b, c). Note that displacements are enlarged, and displacement of X atom in the b-direction (w) is not shown to avoid confusion (see Fig. 1.3)

Table 1.3. Crystallographic parameters of MnP and CrAs in the MnP-type structure (room temperature) [4]

	δ	u	v	x	w
MnP	0.077	0.05	0.06	0.005	0.01
CrAs	0.036	0.05	0.05	0.007	0.006

Table 1.4. Positions of atoms in the MnP-type structure expressed in the coordinates based on *A*-, *B*-, and *C*-axes

M	$(-x,0,u),(x,1/2,1/2-u),\ (1/2-x,0,-u),\ (1/2+x,1/2,1/2+u)$
X	$(1/4-v,1/2,5/6-w),\ (1/4+v,0,1/3-w),$ $(3/4-v,1/2,1/6+w),\ (3/4+v,0,2/3+w)$

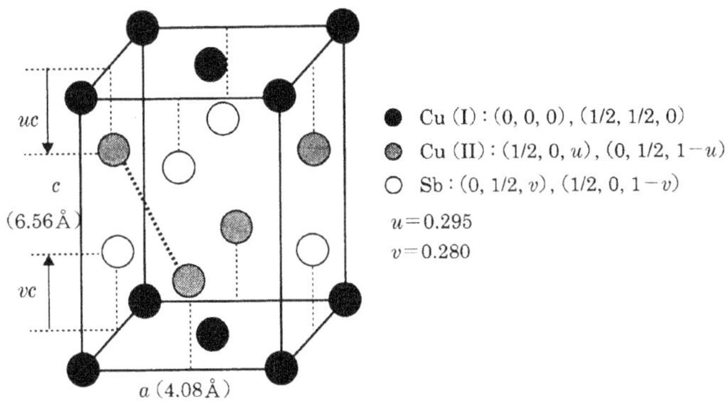

Fig. 1.5. Unit cell of the Cu_2Sb-type structure. 3*d* transition metal occupies the Cu(I)- and Cu(II)-sites. In MnAlGe, Al atoms occupy the Cu(II)-sites and Ge occupies the position of Sb. The crystallographic parameters *u* and *v* are these of Mn_2Sb

reliable accuracy by X-ray diffraction for powdered samples. Table 1.4 shows position of atoms (see Figs. 1.3 and 1.4). We note that the coordinates are based on the *A*-, *B*-, and *C*-axes (*c*-, *a*-, and *b*-axes) shown in Fig. 1.4.

In some literatures, origin of coordinate for the atomic position is shifted to $(1/4,1/4)$ in the *ab* (*BC*)-plane shown in Fig. 1.3 (or 1.4). In this case, the positions shown in Table 1.4 must also be shifted. There have been several (about four) manners of expression for the lattice parameters of the MnP-type structure. The (a,b,c)-system used in Figs. 1.3 and 1.4 is useful because the *a*- and *c*-axes of the NiAs-type lattice correspond to the *a*- and *c*-axes of the MnP-type structure. We use mainly (a,b,c) system and sometimes (A,B,C) system in this book.

1.3.3 Cu_2Sb-Type Structure

MnAlGe, MnGaGe, and M_2X denoted by boldface letters in Table 1.1 take this crystal structure, whose unit cell is shown in Fig. 1.5. In this structure, distance between the Mn(II) atoms (denoted by the thick dotted line) is large. For example, the distance is 3.94 Å for Mn_2Sb. Nearest distance between Mn(I)–Mn(II) and Mn(I)–Mn(I) is 2.81 and 2.88 Å, respectively. In the case of MnAlGe, which is obtained by substituting Mn(II) and Sb of Mn_2Sb by Al and Ge, respectively, Mn–Mn distance along the c-axis is very large and equal to c (= 5.93 Å). This is also the case of MnGaGe. Crystallographic characteristics mentioned above affects magnetic properties, which we discuss in Chap. 3.

1.4 Nonstoichiometric Compounds

One of typical nonstoichiometric MX compounds is $Mn_{1+x}Sb$. Figure 1.6 is a phase diagram of Mn–Sb binary system [3]. The NiAs-type structure exists at room temperature for Sb composition between 44% and 49%. In other words, $Mn_{1+x}Sb$ with $0.04 \leq x \leq 0.25$ has the NiAs-type structure. Curie temperature T_C of $Mn_{1+x}Sb$ decreases linearly from 314°C (51 at% Mn) to 90°C (56 at% Mn) with increasing Mn composition. This diagram shows that stoichiometric (i.e., 50 at% Mn) MnSb does not exist. MnSb is considered to have about 1% excessive Mn atoms occupying partially the interstitial sites ((2/3, 1/3, 1/4) and (1/3, 2/3, 3/4)) shown in Fig. 1.7. It has been reported that an excessive Mn atom at the interstitial site is surrounded by six nearest Mn atoms and does not have magnetic moment, while an Sb atom polarizes with small moment $0.2\mu_B$ antiparallel to the total moment [5]. For $Mn_{1+x}Sb$ ($0.04 \leq x \leq 0.25$), T_C and spontaneous magnetization rapidly decrease with increasing x. Magnetic anisotropy constants also vary with x [6].

Some other MX compounds that forms the NiAs- (or MnP-) type structure in nonstoichiometric composition region are shown in Table 1.5.

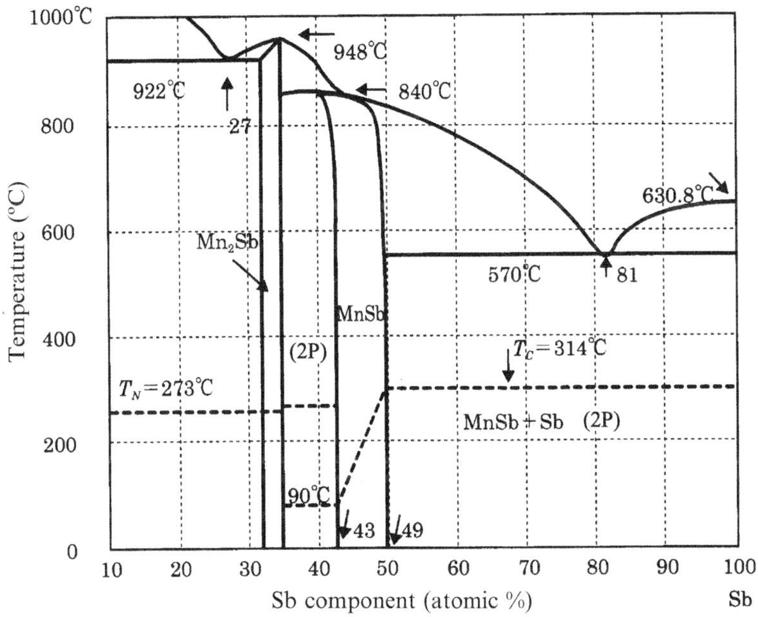

Fig. 1.6. A phase diagram of the Mn–Sb binary system. The label 2P denotes two phases

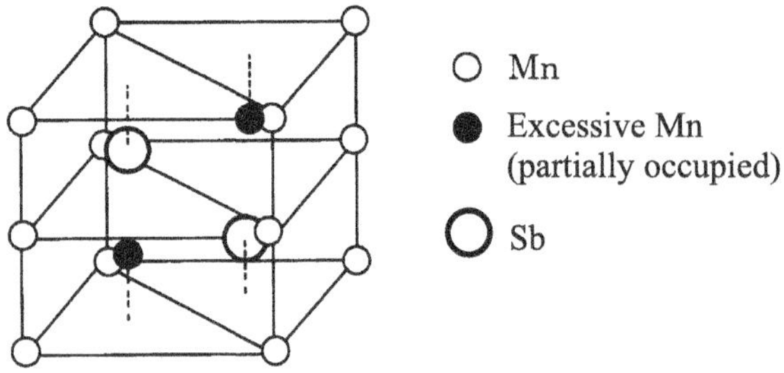

Fig. 1.7. Interstitial positions of the NiAs-type structure at which excessive Mn atoms of $Mn_{1+x}Sb$ locate are denoted by the *filled circles*

Table 1.5. Composition region where MX-type compound exists (room temperature)

VSb	Single phase only for $x = 0.4$ [7]
CrSb	$0 \leq x \leq 0.04$. Upper limit of x increases with temp [1,3]
MnSb	See Fig. 1.6
FeSb	$0.2 \leq x \leq 0.3$ [8], $0.1 \leq x \leq 0.3$ [3] Region incr. with temp [8] Exists only for excessive Fe
CoSb	$0 \leq x \leq 0.04$. Region incr. with temp [1]. $-0.04 \leq x \leq 0.15$ [3]
NiSb	$-0.06 \leq x \leq 0.02$. Region incr. with temp [1]. $-0.08 \leq x \leq 0.10$ [3]
VAs	1:1 compound
VP	1:1 compound
CrP	1:1 compound
CrAs	1:1 compound
MnP	1:1 compound
MnAs	1:1 compound
MnBi	MnBi for $T < 340°C$. $Mn_{1.08}Bi$ for $T > 340°C$ Phase separation into Mn and Mn-Bi liquid for $T > 446°C$ [3,8]

The letter x in this table means composition in $M_{1+x}X$. Generally, the region expands as temperature rises

References

[1] A. Kjekshus, K. P. Walseth, Acta Chem. Scand. **23**, 2621 (1969)
[2] T. Okita, Y. Makino, J. Phys. Soc. Jpn. **25**, 120 (1968)
[3] H. Okamoto (ed.), *Phase Diagram for Binary Alloys*, (ASM International, OH, 2000)
[4] Landolt-Börnstein III/27a, *Magnetic Properties of Pnictides and Chalcogenides*, ed. by K. Adachi, S. Ogawa (Springer, Berlin, 1989), p. 70
[5] Y. Yamaguchi, H. Watanabe, T. Suzuki, J.Phys. Soc. Jpn. **45**, 846 (1978)
[6] T. Okita, Y. Makino, J.Phys. Soc. Jpn. **25**, 120 (1968)
[7] J. Bouma, C.F. van Bruggen, C. Haas, J. Solid State Chem. **7**, 255 (1973)
[8] T. Chen, W. Stutius, IEEE Trans. Magn. **10**, 581 (1974)

2

Overview of Magnetic Properties of NiAs-Type (MnP-Type) and Cu_2Sb-Type Compounds

2.1 Compounds That Have Magnetic Ordering Phase

In this section, we look over structural and magnetic properties of the MX and M_2X compounds. In Table 2.1, we list the NiAs- (or MnP-) type and the Cu_2Sb-type compounds that show magnetic ordering. Basic magnetic properties of typical MX and M_2X compounds are shown in Tables 2.2–2.6. In the tables, only representative data are shown when there are several data. Refer to Chap. 3 for more detailed magnetic properties.

As shown in Tables 2.1–2.4, magnetic ordering occurs only in compounds containing Cr, Mn or Fe. The compounds with the MnP-type structure show the double helical magnetic ordering as shown in Table 2.6.

As a typical example of the double helical structure, we explain the magnetic structure of CrAs in Fig. 2.1.

We summarize the double helical magnetic orderings of the MnP-type compounds in Table 2.6. MnP has a long wavelength and thus is close to a ferromagnet, and is known to show metamagnetic magnetization process. Magnetization of MnP saturates at magnetic field about 4 kOe when the applied field is along the *a*-axis. There are no magnetization measurements for the other compounds listed in Table 2.6. There is only an unpublished data by us for powdered sample of CrAs measured in pulsed fields up to 35 T at liquid nitrogen temperature. According to this report, magnetization shows no saturation and reveals anomaly at 10 T. CrAs is an interesting compound with various unsettled problems. We will discuss some of these problems in Chap. 3.

MnX (X = As, Sb, or Bi) are ferromagnets. CrAs, CrSb, MnAs, and MnBi show a first order phase transition at T_C (or T_N). See Chap. 3 for the detail of these phase transitions.

The Cu_2Sb (M_2X)-type compounds have rich variety of magnetic structures as shown in Fig. 2.2 [1]. In particular, ferrimagnetic Mn_2Sb is interesting; Slight

Table 2.1. The shape of the signs indicates crystal structure at room temperature

X \ M		Ti	V	Cr	Mn	Fe	Co	Ni
P	MP		⬡	⬡	⬢	⬢	⬡	
As	MAs		⬡	⬢	⎔ (gray)	⬢	⬡	⎔
	M_2As			■	■	■		
Sb	MSb	⎔	⎔	⎔ (gray)	⎔ (gray)	⎔ (gray)	⎔	⎔
	M_2Sb				■			
Bi	MBi				⎔ (gray)			

⎔: NiAs-type ⬡: MnP-type □: Cu_2Sb-type

Compounds that show magnetic ordering are denoted by the gray sign

Table 2.2. Magnetic properties of CrX [1]

MX	CrP	CrAs	CrSb
Magnetic structure	χ has a peak ~ 200 K	Double helix	Antiferro
T_N(K)		250 (1st)	718 (1st)
p_A (μ_B/Cr)	$\chi(RT \simeq 3) \times 10^{-6}$	1.67	3.0

Unit of magnetic susceptibility χ is emu g^{-1}. See also Chap. 3. T_N and p_A are Néel temperature and magnetic moment per atom. (1st) means first order transition

Table 2.3. Magnetic properties of MnX [1]

MX	MnP	MnAs	MnSb	MnBi
Magnetic structure	Double helix ($T \leq 47$ K) Ferro ($47\,K \leq T < 291$ K)	Ferro	Ferro	Ferro
T_C (K)	291	318 (1st)	587	628 (1st)
p_A (μ_B/Mn)	1.3	3.4	3.6	3.8

See also Chap. 3. T_C is Curie temperature. (1st) means first order transition

substitution of Cr etc. for Mn induces a first order phase transition from an antiferromagnetic to a ferrimagnetic state with increasing temperature. We will discuss this phenomenon in Chap. 3. The various magnetic ordering shown in Fig. 2.2 is also discussed in Part II of this book.

Table 2.4. Magnetic properties of FeX [1]

MX	FeP	FeAs	$Fe_{1+x}Sb$
Magnetic structure	Double helix	Double helix	Triangular in c-plane
T_N(K)	125	77	$105 \sim 211$ (depends on x)
p_A (μ_B/Fe)	0.41	0.51	0.88

Table 2.5. Magnetic properties of M_2X (See [1] and Chap. 3)

M_2X	Cr_2As	Mn_2As	Fe_2As	Mn_2Sb
Magnetic structure	Antiferro	Antiferro	Antiferro	Ferri
T_N(K)	393	573	353	550
p_A (μ_B/M(I),M(II))	0.40, 1.34	2.2, 4.1	1.28, 2.05	2.13, 3.87

Table 2.6. Double helical structure

	MnP	CrAs	FeP	FeAs
θ_{12}	16°	−126°	169°	140°
$2\pi/k(=\lambda)$	8.2b	2.67b	5.0b	2.67b

See Fig. 2.1 for θ_{12}. b is length of the b-axis. Data are taken at 4 K (12 K for FeAs) (see [1] and Chap. 3)

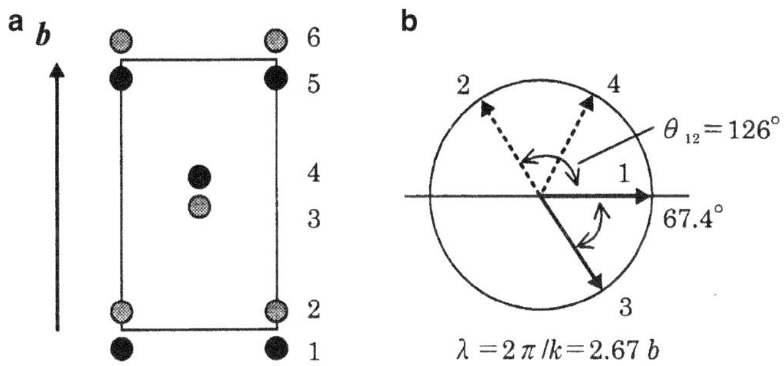

Fig. 2.1. Double helical magnetic ordering of CrAs. (**a**) Cr atoms projected onto the ab-plane (see Fig. 1.3b). The spins are parallel within the plane 1, 2, 3,... normal to the b-axis. (**b**) As the plane changes 1, 3, 5,... (or 2, 4, 6,...), the direction of the spin rotates by 67.4° with the relative angle 126° between adjacent (i.e., between 1 and 2) planes. The wavelength is $\lambda = 2.67b$ and $\mu_{Cr} = 1.67\mu_B$ [2]. Slightly different data are reported in [3]. See also Fig. 4.33 in Part II

2.2 Compounds without Magnetic Ordering

The compounds denoted by unshaded symbols in Table 2.1, which do not include Cr, Mn, or Fe, are paramagnetic or diamagnetic. We summarize properties of these compounds that show no magnetic ordering in Table 2.7.

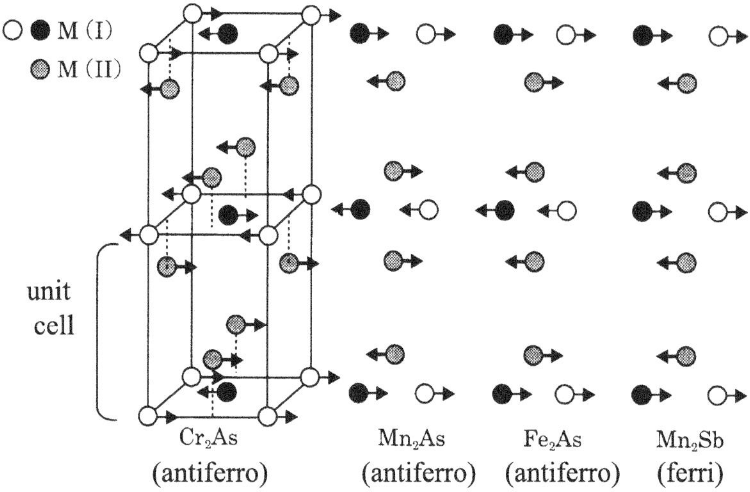

Fig. 2.2. Magnetic orders of the Cu_2Sb-type compounds. M(I) and M(II) (see Fig. 1.5) have different moments. As (Sb) is not shown [1]

Table 2.7. Properties MX compounds that have no magnetic ordering

MX	Struct.	Magnetism (unit of χ is emu g^{-1})	
TiP	TiP		
TiAs	TiP	Pauli para. $\chi \simeq 1.5 \times 10^{-6}$ ($T = 4.2$ K)	[4]
TiSb	NiAs	Pauli para	[5]
VP	NiAs	χ slightly decreases with temp. $\chi \simeq 2 \times 10^{-6}$ (RT)	[6]
VAs	MnP	χ slightly increases with temp. $\chi \simeq 1.5 \times 10^{-6}$ (RT)	[6]
VSb	NiAs	Single phase in $V_{1.4}Sb$. Pauli para. $\chi \simeq 1.5 \times 10^{-6}$	[7]
CrP	MnP	χ has a broad valley at 200 K. $\chi \simeq 3 \times 10^{-6}$ (RT)	[8]
CoP	MnP	Pauli para. $\chi \simeq 1.5 \times 10^{-6}$ ($T \geq 100$ K)	[9]
CoAs	NiAs	χ has a broad peak at 200 K. $\chi \simeq 2 \times 10^{-6}$ (RT)	[10]
CoSb	NiAs	Pauli para. $\chi \simeq 1 \times 10^{-6}$	[4]
NiP	NiP		
NiAs	NiAs	Pauli para. $\chi \simeq 2.5 \times 10^{-7}$ ($T \geq 300$ K)	[11]
NiSb	NiAs	Diamag	[12]
NiBi	NiAs	Superconducting, $T_C = 4.25$ K	[13]

MX compounds that have no magnetic ordering are of little interest in magnetism. However, we can investigate genuine electronic structures that are helpful to understand properties of other MX compounds. In Table 2.7, some of such compounds with weak temperature dependence of susceptibility are denoted by "Pauli para". The susceptibility χ of the order of 10^{-6} to 10^{-7} emu g^{-1} may be considered by not only 3*d* states but also diamagnetism of anion, and possibly by some kind of magnetic impurities. Some compounds such as VSb has a nonstoichiometric phase in which the composition deviates from 1:1. Magnetic properties are often affected by nonstoichiometry. As already described in Sect. 1.4, ferromagnetic $Mn_{1+x}Sb$ is a typical example of nonstoichiometric compounds [7, 14, 15].

References

[1] Landolt-Börnstein III/27a, *Magnetic Properties of Pnictides and Chalcogenides*, ed. by K. Adachi, S. Ogawa (Springer Berlin, 1989), p. 70 and references therein

[2] G.P. Felcher, F.A. Smith, D. Bellavance, A. Wold, Phys. Rev. B **9**, 3046 (1971)

[3] K. Selte, A. Kjekshus, W.A. Jamison, A.F. Andresen, J.E. Engebresen, Acta Chem. Scand. **35**, 1042 (1971)

[4] H. Ido, J.Appl. Phys. part IIA 3247 (1985)

[5] K. Adachi, J. Phys. Soc. Jpn. **16**, 2187 (1961)

[6] K. Selte, A. Kjekshus, A.F. Andresen, Acta Chem. Scand. **26**, 4057 (1972)

[7] J. Bouma, C.F. van Bruggen, C. Haas, J. Solid State Chem. **7**, 255 (1973)

[8] K. Selte, H. Hjersing, A. Kjekshus, A.F. Andresen, P. Fischer, Acta Chem. Scand. A**29**, 695 (1975)

[9] K. Selte, L. Birkeland, A. Kjekshus, Acta Chem. Scand. **A32**, 731 (1978)

[10] K. Selte, A. Kjekshus, Acta Chem. Scand. **25**, 3277 (1971)

[11] I.L.A. Delphin, K. Selte, A. Kjekshus, A.F. Andresen, Acta Chem. Scand. A**32**, 179 (1978)

[12] H. Schmit, Cobalt **7**, 26 (1960)

[13] N.E. Alekseevskii, N.B. Brandt, T.I. Kostina, Izu. Akad. Nauk SSSR **16**, 233 (1952)

[14] Y. Yamaguchi, H. Watanabe, T. Suzuki, J. Phys. Soc. Jpn. **45**, 846 (1978)

[15] T. Okita, Y. Makino, J. Phys. Soc. Jpn. **25**, 120 (1968)

3

Properties of the Compounds with NiAs-Type (MnP-Type) and Cu_2Sb-Type Structures

3.1 MnP and Related Compounds

A magnetic phase diagram of MnP with respect to magnetic field and temperature is shown in Fig. 3.1. This diagram was obtained from magnetization measurements for a single crystal MnP [1]. It has been known that in the temperature region $T \leq 47$ K, MnP shows a double helical ordering with relatively long wavelength of 8.2b propagating along the b-axis [6]. Above 47 K, MnP is a ferromagnet with $T_C = 291$ K. For 420 K $\leq T \leq$ 660 K, susceptibility obeys the Curie–Weiss law with $\mu_{eff} = 2.36\mu_B$ $(2S = 1.56)$ and $\theta_P = 344$ K. This value of μ_{eff} is close to saturation moment $1.3\mu_B$ observed at $T = 0$ K [7]. The χ^{-1}-T curve shows upward convex in the range 660 K $\leq T \leq$ 1400 K [8]. Studies on mixed compounds with the type of $Mn_{1-x}M_xP$ (M = other 3d metal) also have been reported [9–12].

3.2 MnAs and Related Compounds

3.2.1 Magnetic Transition of MnAs and the Bean–Rodbell Theory

MnAs is especially interesting among the MX compounds. In this section we show magnetic and crystallographic properties of MnAs and give a simple explanation for them on the basis of the phenomenological theory developed by Bean and Rodbell. We also mention some interesting properties of mixed compounds between MnAs and MnP (and MnSb).[1]

MnAs has long been of interest since the study carried out by Guillaud et al. [13]. Temperature dependence of magnetization and inverse susceptibility are shown schematically in Fig. 3.2 [14]. With rising temperature, spontaneous magnetization disappears discontinuously at $T_C(\mathrm{up}) = 318$ K. This first order transition is accompanied with structural transformation from the NiAs-type to the MnP-type structure. Susceptibility shows a peak at $T_t = 398$ K and obeys the Curie–Weiss law for $T > T_t$.

[1] See Part II of this book for the itinerant theory on magnetism of MnAs.

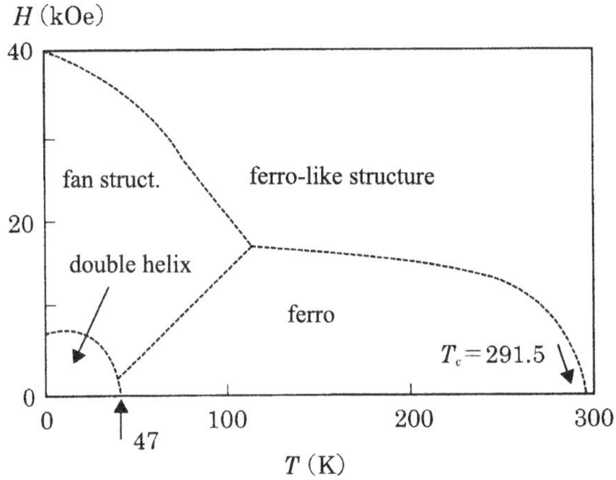

Fig. 3.1. Magnetic phase diagram of MnP [1–4]. See also [5] for more detailed diagram

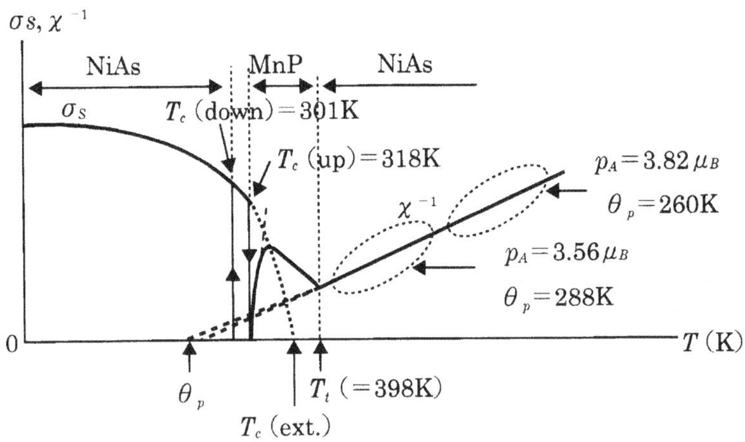

Fig. 3.2. Spontaneous magnetization σ_s and inverse susceptibility χ^{-1} of MnAs plotted as functions of temperature

As we will see later, T_t is a structural transition temperature from the MnP-type to the NiAs-type structure. Anomalous temperature dependence of paramagnetic susceptibility in the temperature range $T_C \leq T \leq T_t$ is explained from the viewpoint of itinerant electrons (see Sect. 4.3).

Figure 3.3 shows temperature dependence of the crystallographic parameters [15]. The a-axis and the unit cell volume show remarkable discontinuous changes of 1.2% and 2.1%, respectively, at T_C(up), while the c-axis does not show visible change at T_C(up). These facts suggest that there is a correlation between the a-axis and ferromagnetism of MnAs. Thermal expansion coefficient has a large value in the MnP-type phase just above T_C. This is due to the rapid structural transformation

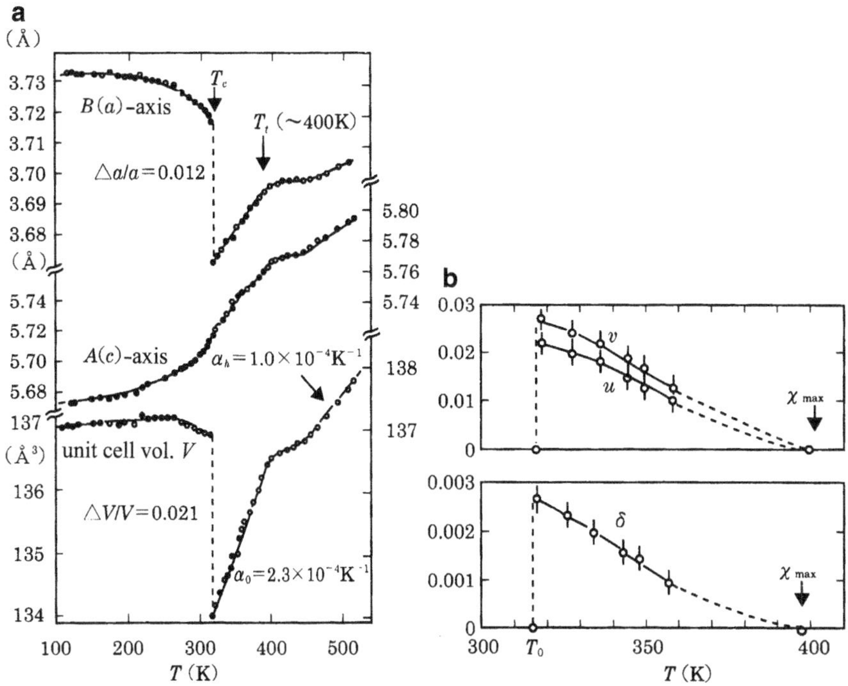

Fig. 3.3. (**a**) Temperature dependence of lattice constants of MnAs. (a, b, c are lattice constants of the hexagonal lattice.) α is the thermal expansion coefficient. (**b**) Temperature dependence of the parameters u, v, and δ. $\chi_{\max}$ means the temperature where the χ^{-1}-T curve shows the minimum as shown in Fig. 3.2 (Reproduced from [15])

from the MnP-type to the NiAs-type in the small range of temperature (see Part II for detail). The crystallographic parameters u, v, and δ (see Table 1.3) take the maximum value just above T_C(up), approaching zero at T_t on further increasing temperature as indicated in Fig. 3.3b.

MnAs shows a metamagnetic magnetization process in the temperature range $T_C < T < T_t$, corresponding to a paramagnetic–ferromagnetic transition accompanied with a crystallographic change from the MnP-type to the NiAs-type structures. As the magnetization process is similar to that of $MnAs_{1-x}P_x$, which we will discuss later, we show in this section only temperature change of magnetization in $H = 2$ T and $H = 40$ T in Fig. 3.4 [16]. The curve in $H = 40$ T suggests that, if the NiAs-type structure below T_C were maintained up to the temperature region just above T_C of the spontaneous magnetization, T_C could be about 450 K. It is noted that this roughly extrapolated Curie temperature $T_C \simeq 450$ K is considerably higher than $\theta_p \simeq 270$ K estimated in the high-temperature NiAs-type region, which is characteristic of MnAs as $T_C \simeq \theta_p$ is satisfied for MnSb and MnP.

Various characteristic properties of MnAs mentioned above are well explained by a phenomenological theory based on an assumption of strong volume dependence

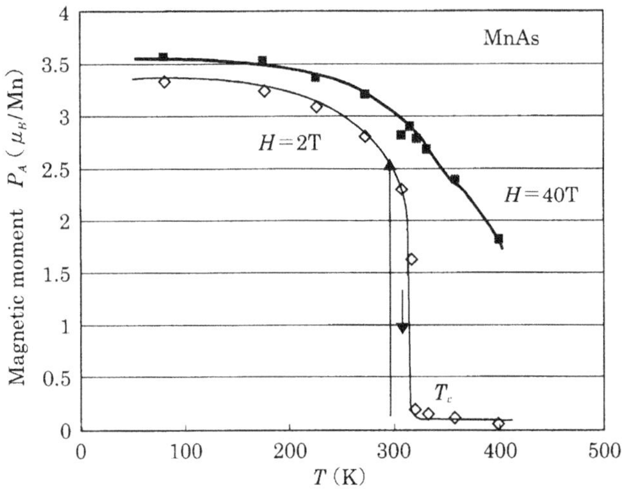

Fig. 3.4. Temperature dependence of magnetization of MnAs in two magnetic fields $H = 2$ and 40 T. T_C means T_C(up) (Reproduced from [16])

of exchange interaction between Mn atoms [17, 18]. Namely, the first order transition of MnAs (such as the discontinuous change of crystal volume and spontaneous magnetization, and their hysteresis for temperature variation), metamagnetic phase transition in the temperature region just above T_C (e.g., magnetization process and temperature dependence of hysteresis), and pressure dependence of T_C can be generally explained. We have to note, however, that 3*d* electrons in MnAs from which magnetization arises have itinerant character. Therefore, it is necessary to treat 3*d* electrons in MnAs as itinerant electrons as we will show in Part II of this book. In this section we describe the phenomenological theory to show how the various peculiar properties of MnAs are explained.

First, we assume that the Curie temperature depends on crystal volume (or lattice constants) [17]. This assumption is reasonable as T_C depends on applied pressure in most ferromagnets. The crystal volume is affected by various factors, for example, thermal expansion, pressure, applied magnetic field, and magnetovolume effect (magnetostriction). Structural phase transition also changes crystal volume. Considering such factors, Bean and Rodbell [17] set a relation

$$T_C(V) = T_0[1 + \beta(V - V_0)/V_0], \tag{3.1}$$

where T_C is assumed to be a linear function of crystal volume V. This may be justified as long as volume change is small. In (3.1), V_0 is the volume at $T = 0$ K in the absence of magnetovolume effects. T_0 is an expected Curie temperature when V is assumed to be kept in the constant V_0. Volume dependence of T_C is given by the coefficient β, which takes a large value for MnAs.

We consider a localized model where a magnetic material consists of N magnetic atoms with spin j. Under an applied magnetic field H and pressure P, free energy of this material is written in the form

$$
\begin{aligned}
G = &-(3/2)[j/(j+1)]Nk_{\mathrm{B}}T_{\mathrm{C}}\sigma^2 - HM_{\mathrm{s}}\sigma + (1/2K)[(V-V_0)/V_0]^2 \\
&- Tk_{\mathrm{B}}NS + P[(V-V_0)/V_0],
\end{aligned}
\tag{3.2}
$$

where $\sigma(=M_{\mathrm{s}}(T)/M_{\mathrm{s}}(0))$ is a magnetization normalized by $M_{\mathrm{S}}(0)$ specific magnetization at 0 K. The first term denotes exchange energy written by using T_{C}.[2] The second and the third terms are Zeeman energy and elastic energy (K is the compressibility), respectively. The fourth term is the entropy one and $k_{\mathrm{B}}NS$ (k_{B} is the Boltzmann constant) is the entropy, where S is summation of magnetic contribution $S_\sigma(\sigma, j)$ and lattice contribution S_{L}. The last term denotes energy due to applied pressure.

When σ is smaller enough than 1, S_σ of the magnetic entropy term $k_{\mathrm{B}}NS_\sigma(\sigma, j)$ can be expanded in the following form:

$$
S_\sigma(\sigma, j) = S_\sigma(0, j) + a\sigma^2 + b\sigma^4 + c\sigma^6 + \ldots,
\tag{3.3}
$$

where the coefficients a, b, ... are given by

$$
a = -(3/2)[j/(j+1)], \quad b = -(9/20)[(2j+1)^4 - 1]/[2(j+1)]^4.
\tag{3.4}
$$

See Appendix for the detailed expression of (3.4). This expansion is useful to investigate magnetic phase transition. We note that the exact expression of $S_\sigma(\sigma, j)$ is given by

$$
S_\sigma(0, j) - S_\sigma(\sigma, j) = \alpha\sigma - \int_0^\alpha B_j(\alpha)\,\mathrm{d}\alpha,
\tag{3.5}
$$

where $B_j(\alpha)$ is the Brillouin function with angular momentum j. For the discussion in this section, we use (3.3) and (3.6), which is derived from (3.5) as given by

$$
\partial S_\sigma/\partial\sigma = -\alpha.
\tag{3.6}
$$

Substituting (3.1) into (3.2), and using a condition of minimum free energy $\partial G/\partial V = 0$, we have a relation

$$
(V-V_0)/V_0 = (3/2)[j^2/j(j+1)]Nk_{\mathrm{B}}KT_0\beta\sigma^2 - PK.
\tag{3.7}
$$

This equation means that magnetovolume effect (or exchange striction) is proportional to σ^2 if $P = 0$.

First, we consider a case of $P = 0$. Inserting (3.7) into (3.2), we obtain an expression of G, which has been minimized with respect to V. For this free energy, we use a condition $\partial G/\partial\sigma = 0$ and (3.6), and then we have a relation among spontaneous magnetization $\sigma(=M_{\mathrm{s}}(T)/M_{\mathrm{s}}(0))$, temperature T, and a magnetic field H as

$$
(gj\mu_{\mathrm{B}}/k_{\mathrm{B}})(H/T_0) = 2a\sigma[1 + (2b/a)\eta\sigma^2] + (T/T_0)\alpha \;\cdots,
\tag{3.8}
$$

[2] Exchange energy is written as $-(1/2)AM_{\mathrm{s}}^2$ with the coefficient A of molecular field. The first term in (3.2) is derived from this expression.

where μ_B is the Bohr magneton and g is the g factor (=2 in the present case). In this equation η is defined by

$$\eta = -(a^2/2b)Nk_B KT_0\beta^2. \tag{3.9}$$

We note that η is proportional to β^2. As we will show below, η is an important parameter contained in the coefficient of the σ^4 term of the free energy expansion. For $\eta > 1$, a first order transition occurs, while order of transition is second for $\eta < 1$ (see also Appendix). The coefficients a and b are given in (3.4), and α is a function of σ (i.e., $\sigma = B_j(\alpha)$).

Using (3.8) with a given value for j and setting $H = 0$, we can calculate temperature variation of spontaneous relative magnetization $\sigma = \sigma(T/T_0)$ for various values of η. We can also calculate magnetization curve by fixing temperature and changing H. By considering that magnetization of MnAs at $T = 0$ K is $3.4\mu_B$, we set $j = 3/2$ (with $g = 2$). Figure 3.5 shows temperature dependence of σ calculated from (3.8) for various values of η. Magnetization curves calculated for $\eta = 2$ at various temperatures are plotted in Fig. 3.6.

We see from Fig. 3.5 that $\eta = 1$ is the critical value between the first and the second order transition. For $\eta = 2$, relative spontaneous magnetization σ is 0.81 at $t = 1$ and is 0.6 at $t = 1.08$ (i.e. just below T_C). These results agree well with experimental values shown in Fig. 3.4. We also see that magnitude of hysteresis of spontaneous magnetization is $\Delta(T/T_0) = 0.08$. This value is also close to the experimental value $\Delta(T/T_0) = 0.06$ (i.e., $T_C(\text{up}) = 318$ K and $T_C(\text{down}) = T_0 = 301$ K). Therefore, temperature dependence of spontaneous magnetization of MnAs is well reproduced with the parameter $\eta = 2$.

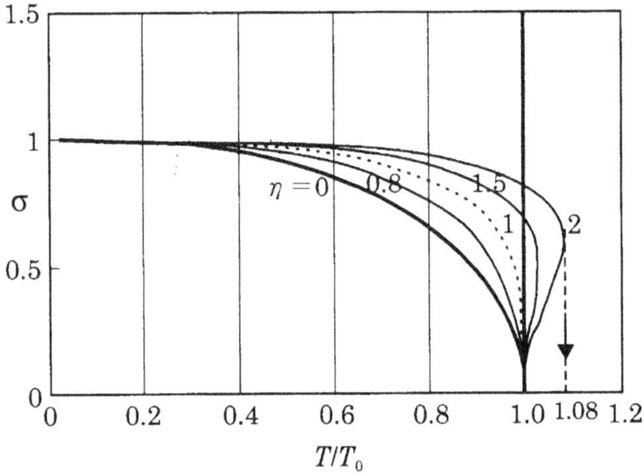

Fig. 3.5. Relation between magnetization (σ) and normalized temperature (T/T_0) for various values of η. The curves were calculated on the basis of (3.8). The curve with $\eta = 2$ is in good agreement with experimental curve for MnAs

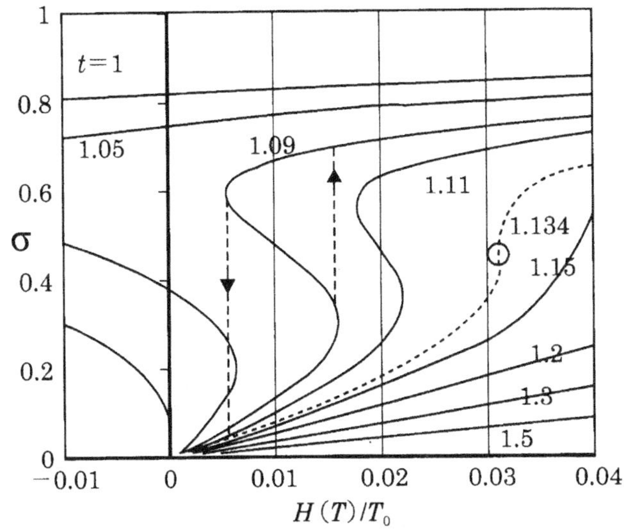

Fig. 3.6. Calculated magnetization curves of MnAs at various temperatures ($T/T_0 = t$) with $\eta = 2$ fixed. We see that spontaneous magnetization is $\sigma = 0.81$ at $t = 1$. The curve at $t = 1.09$ shows metamagnetic behavior clearly but the curves for $t \geq 1.134$ do not show hysteresis as well as magnetization jump

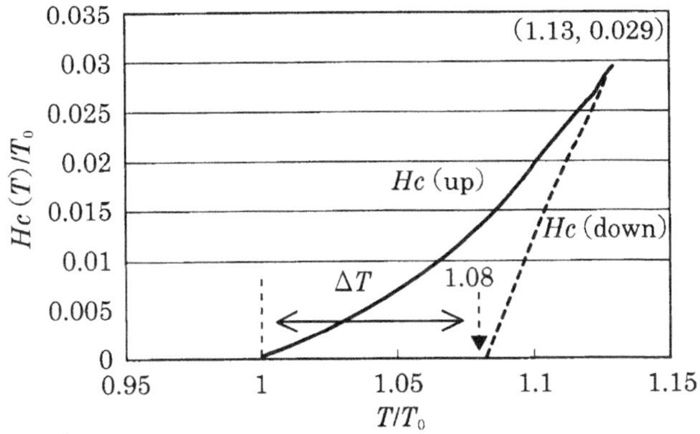

Fig. 3.7. Temperature dependence of H_C(up) and H_C(down) of metamagnetic phase transition calculated for MnAs using $\eta = 2$

It is known that MnAs shows metamagnetic behavior at temperatures higher than T_C(up) [19]. Figure 3.6 is the result calculated from (3.8) with $\eta = 2$. The magnetization curves for $1 < T/T_0 \leq 1.134$ show a jump at the transition field H_C. These curves also show hysteresis as indicated by the vertical dotted lines for $t = 1.09$. Figure 3.7 shows the temperature variation of transition fields for the process of increasing and decreasing field evaluated from Fig. 3.6.

Setting $T_0 = T_C(\text{down}) = 301$ K, the critical point ($T/T_0 = 1.13$ and $H_C/T_0 = 0.029$) in Fig. 3.7 corresponds to $T = 340$ K and $H = 8.7$ T. These values agree well with experimental results $T = 345$ K and $H = 9$ T found by Zieba [20]. Magnetization curves measured by Grazhdankina [19] show that the critical point locates near 350 K, which is also in good agreement with the calculated results mentioned above. We note that Fig. 3.7 is also applicable to metamagnetic behavior of $MnAs_{1-x}P_x$ above T_N shown in the next section.

In summary, anomalous magnetic properties of MnAs is well explained by a phenomenological theory in which exchange interaction between Mn is sensitive to crystal volume (that is, η or β is large).

Next we apply this theory to the magnetovolume effect of MnAs. We start from (3.7). According to experiments, the jump of magnetization at the temperature $T/T_0 = 1.08$ is about 0.6. This value is reproduced with $\eta = 2$. By inserting the value of $T/T_0 = 1.08$ and $\Delta V/V = 0.021$ (obtained from experiments) shown in Fig. 3.3 into (3.7), we have

$$Nk_B KT_0\beta = 0.0648. \tag{3.10}$$

Also, a relation

$$\eta = 2 = 2.21 \times Nk_B KT_0\beta^2 \tag{3.11}$$

is known from temperature dependence of magnetization. From (3.10) and (3.11), we obtain $\beta = 14$ and $K = 3.8 \times 10^{-12}$ (in CGS). Temperature dependence of magnetization as well as magnetovolume effect at $T = T_C(\text{up})$ ($\Delta V/V = 0.021$) is reproduced using these values. However, the experimental value of $\Delta V/V$ is a summation of magnetovolume effect and volume change due to structural transformation from the NiAs-type to the MnP-type. To investigate pure magnetovolume effect, it is necessary to investigate $MnAs_{1-x}Sb_x$, which does not show the structural transformation for $x \geq 0.1$ [21, 22]. We note that the magnetovolume effect of $MnAs_{0.9}Sb_{0.1}$ is well explained with the critical value $\eta = 1$.

Finally, we investigate the effect of pressure on the magnetic transition temperature T_0 of MnAs. In the procedure to derive (3.8), we set $H = 0$ and keep P finite instead of setting $P = 0$. Then we have a relation

$$0 = 2a\sigma[(1-\beta KP) + (2b/a)\eta\sigma^2] + (T/T_0)\alpha. \tag{3.12}$$

This equation shows a relation between σ and T/T_0 under pressure P. We set $\eta = 2$ and $T_0 = 301$ K as we did for MnAs. The value of βK is evaluated to be $\beta K = 5.32 \times 10^{-11}$ from (3.10) and (3.11). Using these values of parameters, we calculated temperature dependence of σ under various pressures as shown in Fig. 3.8. From this figure, $T_C(\text{up})$ and $T_C(\text{down})$ are obtained as a function of pressure as plotted in Fig. 3.9.

The calculated results shown in Fig. 3.9 agree qualitatively with the pressure dependence of $T_C(\text{up})$ and $T_C(\text{down})$ in the low pressure region in Fig. 3.10. We see from this figure that MnAs under pressure transforms easily to the MnP-type structure. We also see that this transformation is of the first order accompanied with large

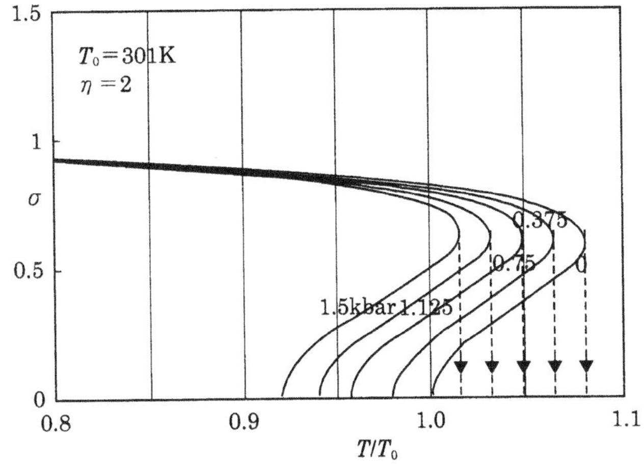

Fig. 3.8. σ of MnAs under pressure as a function of temperature calculated using $K\beta = 5.3 \times 10^{-11}$(CGS)

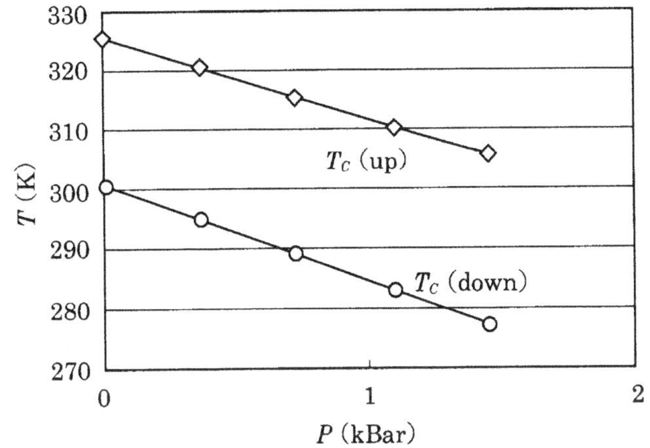

Fig. 3.9. Pressure dependence of first order transition temperatures T_C(up) and T_C(down) of MnAs evaluated from Fig. 3.8

hysteresis for pressure. For example, if we apply pressure larger than $P = 3.1$ kbar (0.31 GP) at $T = 100$ K, the crystal remains in the MnP-type even after the pressure is removed. This phase diagram is similar to that of $MnAs_{1-x}P_x$ (Fig. 3.12), which we will discuss in the next section. These results suggest that NiAs-type structure of ferromagnetic MnAs is quite unstable; it seems to be maintained by occurrence of the ferromagnetic order.

The model we have shown in this section is based on assumptions that Mn atom in MnAs has a local spin 3/2 and that the exchange interaction depends strongly and linearly on the crystal volume. This phenomenological model has explained

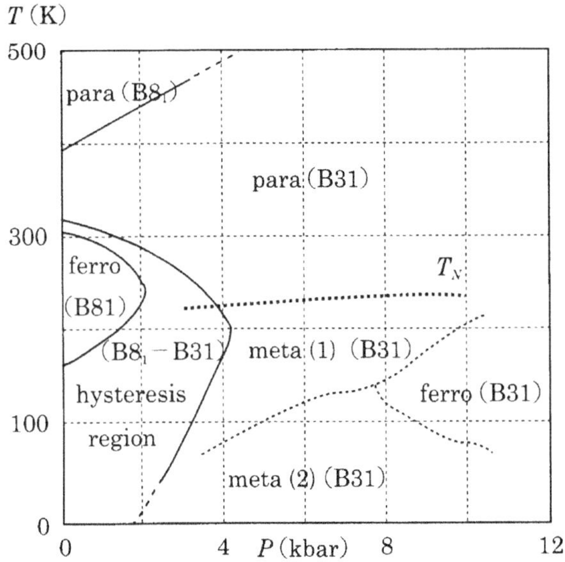

Fig. 3.10. A phase diagram (pressure–temperature) of MnAs (Reproduced from [23])

successfully characteristic properties of MnAs. This model, however, has defects. One is that it is based on the localized spin model. Another is that the structural transformation at T_C(up) is neglected. In Part II of this book, we will show that the anomalous magnetic properties and structural transformation of MnAs are well explained from the viewpoint of itinerant electrons and band structures.

3.2.2 Various Type of Phase Transition of $MnAs_{1-x}P_x$

There are various interesting mixed compounds of MnAs such as $Mn_{1-x}Cr_xAs$ [24–27], $MnAs_{1-x}P_x$ [15,28–30], and $MnAs_{1-x}Sb_x$ (see Sect. 3.2.3) [5]. As we have noted, MnAs is unstable against the MnP-type lattice distortion. Substituting P for As realizes the MnP-type structure. On the other hand, by substitution of Sb for As, the NiAs-type structure is stabilized. We describe $MnAs_{1-x}P_x$ in this section.

First we show a relation between magnetic moment and the a-axis of MnX (X = P, As, Sb, or Bi) in Fig. 3.11. In this figure, MnP and $MnAs_{0.88}P_{0.12}$ takes the MnP-type structure. (We note that the a-axis of the MnP-type structure in this figure corresponds to that of the NiAs-type structure.) We see that magnetic moment is sensitive to P-substitution, but the moment does not change largely as long as the a-axis is longer than that of MnAs.

Figure 3.12 shows a phase diagram of the MnAs–MnP system, which is depicted using data in [31–34]. In this figure, the hatched region indicates the region in which the NiAs-type structure realizes. The compound in the vicinity of MnAs ($x \leq 0.03$) is a ferromagnet, which shows a first order transition at T_C. In the region of the MnP-phase, the double helical ordering realizes although thorough investigation has not been performed yet. The first investigation of these mixed compounds

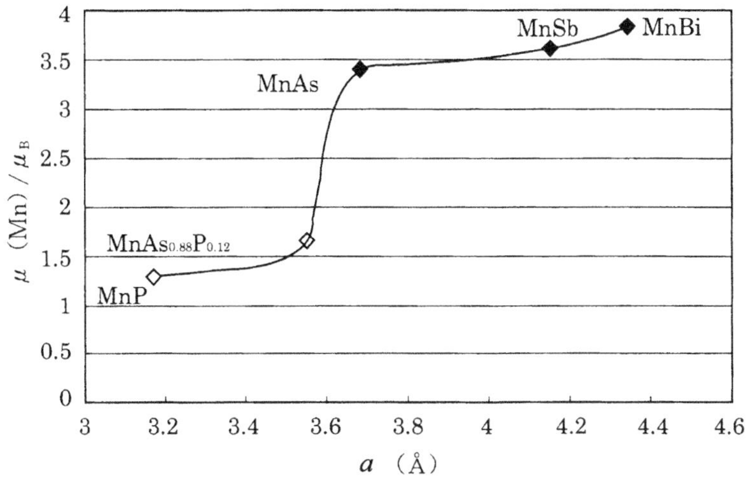

Fig. 3.11. Relation between magnetic moment and the *a*-axis of MnX compounds. The NiAs-type and the MnP-type structure are denoted by *filled* and *open* symbols, respectively

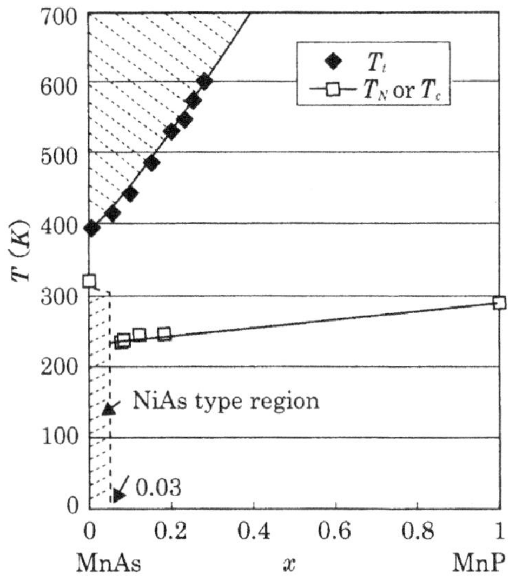

Fig. 3.12. A phase diagram (magnetic state and crystal structure) of the MnAs–MnP mixed crystal

were carried out by Goodenough et al. [32]. In Fig. 3.13, we show *M*-*T* curves of $MnAs_{0.9}P_{0.1}$. The Néel temperature is about 230 K. At the temperatures $T \leq 100\,K$ the double helical ordering realizes. By applying a magnetic field, the double helical structure transforms into ferromagnetic ordering with moment $1.6\mu_B/Mn$ [35]. In the temperature range $100\,K \leq T \leq T_N$, magnetic structure is unknown though

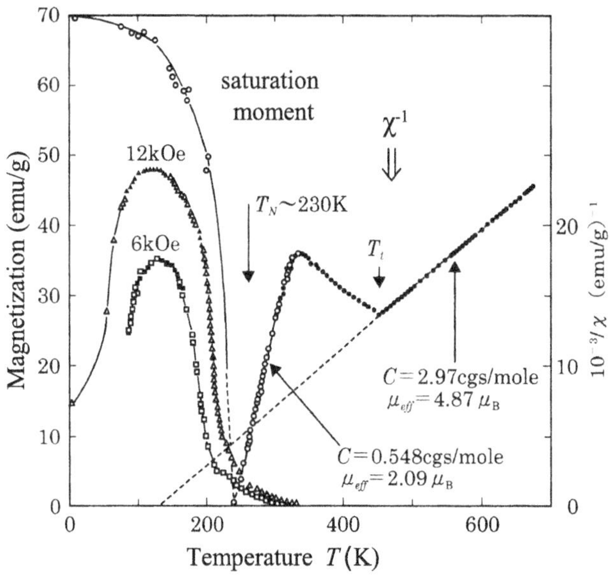

Fig. 3.13. Magnetic moment [35] and inverse susceptibility of $MnAs_{0.9}P_{0.1}$ plotted against temperature. μ_{eff} is effective Bohr magneton (Reproduced from [32])

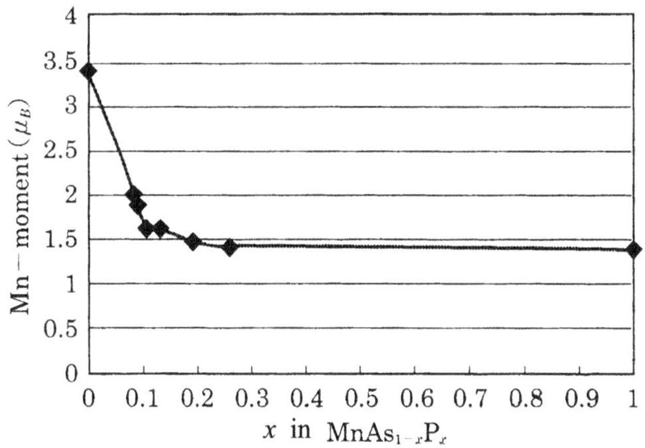

Fig. 3.14. Magnetic moment of Mn in $MnAs_{1-x}P_x$. The data for $x = 0$, 0.1, 0.25, and 1 were obtained from saturation moment measurement [14, 35–37]. Others were taken from results of neutron diffraction measurements [28, 31]

ferromagnetic behavior of a magnetization curve has been observed. Magnetic moment of $MnAs_{1-x}P_x$ was measured by neutron diffraction [28, 31] and saturation moment measurements [16, 35]. The moment slightly decreases with increasing P component for $x \geq 0.03$ as shown in Fig. 3.14.

As we see from Fig. 3.13, behavior of χ^{-1} above T_N is similar to that of MnAs above T_C(up). In the temperature range where χ^{-1} decreases with rising temperature, the MnP-type structure is changing into the NiAs-type (the parameters u, v, and δ approach zero) [15, 36], and the NiAs-type structure realizes above T_t. Moment estimated from the Curie–Weiss-like behavior of χ^{-1} is $\mu_{eff} = 2.09\mu_B (p_A = gS\mu_B = 1.3\mu_B)$. This is close to the saturation moment observed at low temperatures shown in Fig. 3.14. In the range $T \geq T_t$, μ_{eff} is $4.87\mu_B$, indicating that magnetic moment is as large as the moment of MnAs.

Figure 3.13 suggests that transition between a low spin state and a high spin state occurs accompanied with the structural transformation. This transition was discussed first by Goodenough et al. [32]. There is an experiment that indicates relation between susceptibility and crystal structures [15, 36]. Such problems have been explained from the viewpoint of itinerant electrons by Motizuki and her co-workers as described in Part II of this book.

Next we describe field-induced structural transformation of $MnAs_{0.9}P_{0.1}$. We show temperature dependence of moment of $MnAs_{0.9}P_{0.1}$ in high magnetic fields in Fig. 3.15 [16]. As shown in Fig. 3.16, metamagnetic behavior is seen above T_N. This is similar to the behavior of moment of MnAs just above T_C(up). The data shown in Figs. 3.15 and 3.16 were measured at the high magnetic facility of Osaka University. Behavior similar to Figs. 3.15 and 3.16 have been observed for $Mn_{0.9}Cr_{0.9}As$ [37].

In a moderate magnetic field, $MnAs_{0.9}P_{0.1}$ has the double helical ordering with low spin state below $T_N \simeq 230$ K. In high-fields, metamagnetic transition to the state similar to MnAs occurs for $T \geq T_N$ as shown in Figs. 3.15 and 3.16. This

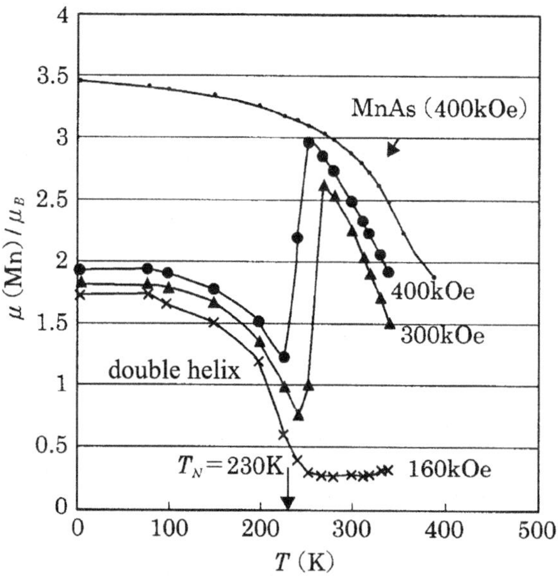

Fig. 3.15. Temperature change of magnetization of $MnAs_{0.9}P_{0.1}$ under high magnetic fields. Data of MnAs are also plotted for comparison (Reproduced from [16])

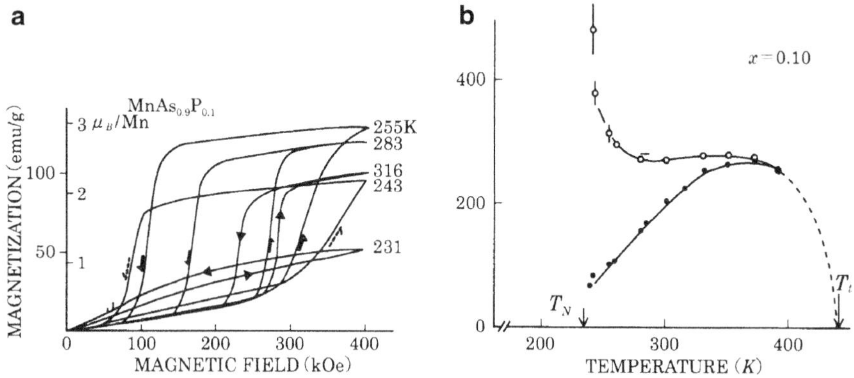

Fig. 3.16. Magnetization process (**a**) and transition fields (**b**) of metamagnetic behavior of $MnAs_{0.9}P_{0.1}$. The *open* (*filled*) symbols are for increasing (decreasing) field (Reproduced from [16])

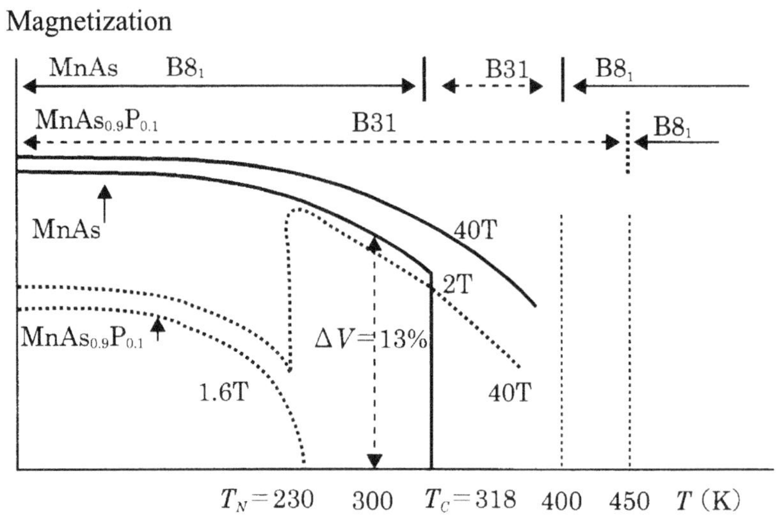

Fig. 3.17. Schematic figure for magnetization–temperature curves of MnAs and $MnAs_{0.9}P_{0.1}$ under high magnetic fields. Temperature range for the NiAs-type and MnP-type crystal structure is also shown

field-induced phase transition is expected to be a transition from paramagnetic state (MnP-type structure, the low spin state) to ferromagnetic state (NiAs-type, the high spin state). This transition is similar to the metamagnetic transition of MnAs just above T_C, as we have noted. Figure 3.17 shows a comparison between MnAs and $MnAs_{0.9}P_{0.1}$.

The fact that the magnetization of $MnAs_{0.9}P_{0.1}$ is almost the same as that of MnAs in the temperature region between T_N (=250 K) and T_C (=318 K), as well

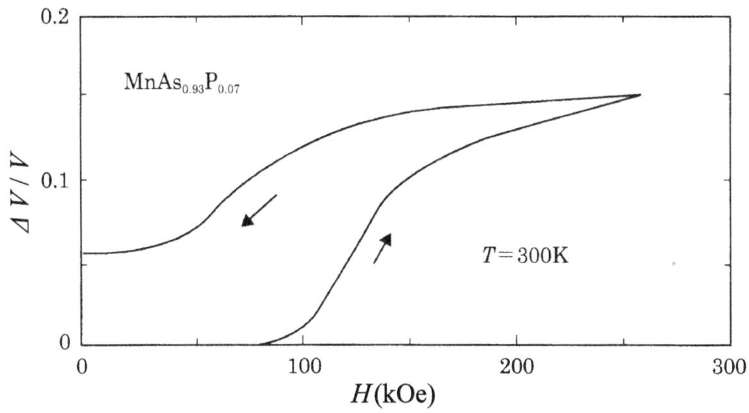

Fig. 3.18. Magnetostriction of $MnAs_{0.93}P_{0.07}$ measured using a pulse field at $T = 300$ K [38]

as very large magnetostriction, which will be explored next, indicates that the metamagnetic transition of $MnAs_{0.9}P_{0.1}$ above T_N accompanies the structural transformation. In $MnAs_{0.93}P_{0.07}$, very large magnetovolume effect $\Delta V/V = 0.15$ accompanied with the metamagnetic transition has been observed at $T = 300$ K, as shown in Fig. 3.18 [38].

The magnetovolume effect shown in Fig. 3.18 is extraordinarily large but reasonably understood. Temperature dependence of lattice constants of $MnAs_{1-x}P_x$ is plotted in Fig. 3.19 [15]. Unit cell volume $V(= abc)$ for $x = 0.1$ at 300 K is 121 Å^3, while that of MnAs in the NiAs-type phase is 137 Å^3 at 300 K (Fig. 3.3). Thus, there is 13% difference (as also shown in Fig. 3.17), which suggests that the large magnetovolume effect shown in Fig. 3.18 is reasonable. Therefore, the metamagnetic transition shown in Fig. 3.16 is accompanied with the structural transformation (between the MnP-type and NiAs-type) with extraordinary volume change more than 10%. Metamagnetic transition accompanying large magnetostriction is possible to occur, because increase of elastic energy due to strain is small in the present case (this is because the very large thermal expansion coefficient of $MnAs_{1-x}P_x$ results in a small elastic coefficient). In Fig. 3.16, we see that transition field increases with decreasing temperature, which is closely related to the decrease of thermal expansion coefficient as seen in Fig. 3.19 [15]. The thermal expansion coefficient of $MnAs_{0.9}P_{0.1}$ is small in the temperature range $T < T_N$ and the transition field of the metamagnetic transition shown in Fig. 3.16 shows rapid increase. At $T = 231$ K, metamagnetic transition does not occur even in 400 kOe. Results shown in Fig. 3.16 suggests that no metamagnetic transition occurs for $T \leq T_N$. These facts seems to be important hints to understand the field-induced metamagnetic transition, which occurs only above T_N. The phenomenological theory of metamagnetism of MnAs just above T_C shown in Sect. 3.2.1 may be applicable to $MnAs_{0.9}P_{0.1}$. Divergence of transition field at T_N (Fig. 3.16) and metamagnetic transition of $MnAs_{0.9}P_{0.1}$ accompanied with structural change are complicated and interesting phenomena. Theoretical studies based on the

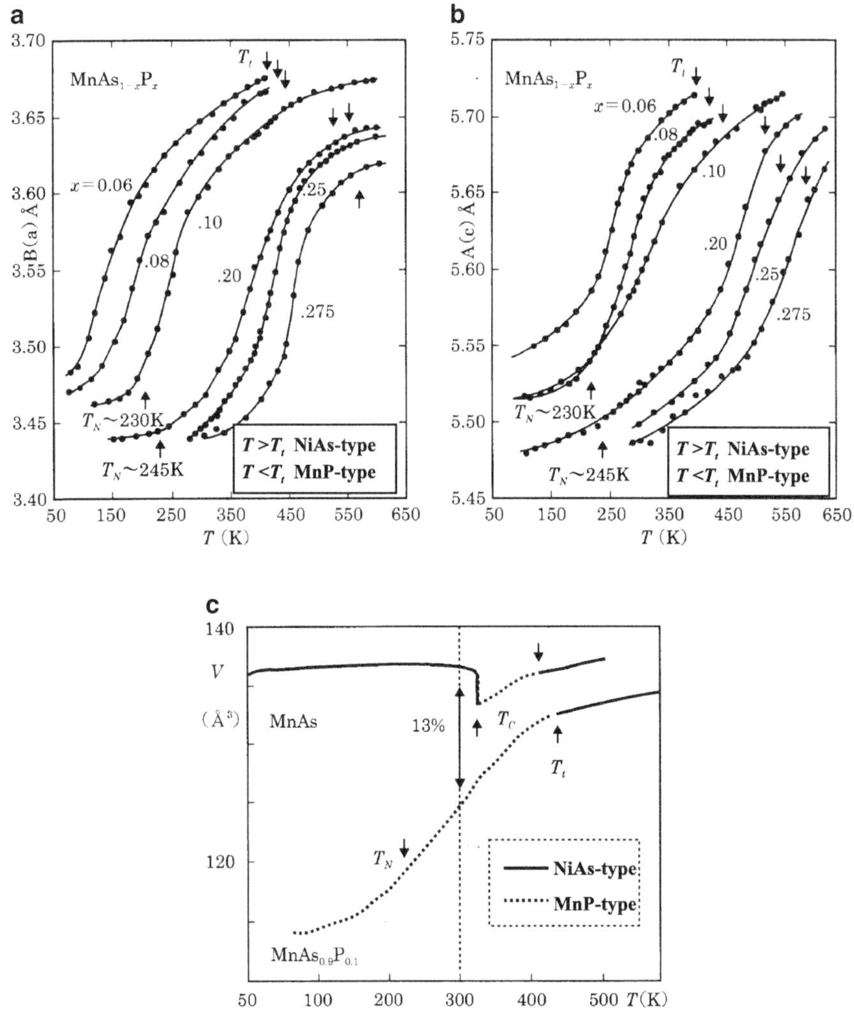

Fig. 3.19. (**a**), (**b**) Lattice parameters of $MnAs_{1-x}P_x$. a and c are the lattice constants of the hexagonal lattice. (**c**) Unit cell volume $V = abc$ of $MnAs_{0.9}P_{0.1}$ and MnAs is schematically plotted as a function of temperature (Reproduced from [15])

itinerant electron picture, as well as experiments in an ultra high field, are necessary to clarify the mechanism of metamagnetic transition in $MnAs_{1-x}P_x$.

3.2.3 Anomalous Behavior of $MnAs_{1-x}Sb_x$

The NiAs-type structure is stabilized by substituting Sb for As of MnAs. Figure 3.20 shows a magnetic and structural phase diagram of the MnAs–MnSb system [39, 40]. The shaded area that indicates the MnP-phase disappears for $x \simeq 0.1$ of $MnAs_{1-x}Sb_x$.

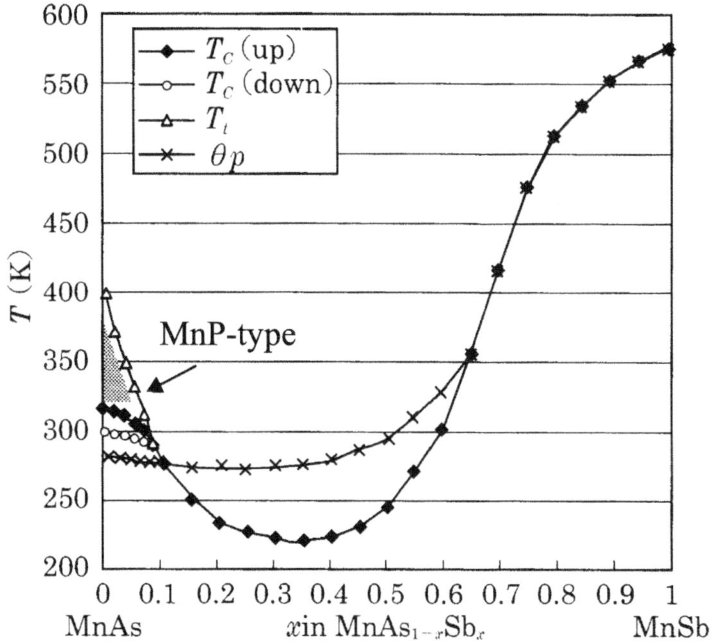

Fig. 3.20. A phase diagram of $MnAs_{1-x}Sb_x$. The shaded area denotes the MnP-type structure. The compound takes the NiAs-type structure in the other area. Ferromagnetism realizes for all composition. For $x < 0.1$, a first order transition occurs at T_C

For $x > 0.1$, magnetic phase transition at T_C is of the second order. T_C takes the minimum value for $x = 0.35$. It is also interesting that T_C and θ_p are significantly different in composition region $0.1 \leq x \leq 0.7$. Magnetic moment per Mn shows weak dependence on x [41] but almost constant in the range $0.15 \leq x \leq 0.7$, where the minimum of T_C ($x = 0.35$) occurs. We will discuss the minimum of T_C later.

As shown in Fig. 3.21, the a-axis increases drastically with Sb composition. The difference of the a-axis between MnAs and MnSb is 13%, while variation of the c-axis is much smaller (about 1% between MnAs and MnSb). The small jumps of lattice parameters at the compositions designated by the vertical dotted lines are explained by the composition dependence of T_C shown in Fig. 3.20. If T_C is higher than the room temperature, the lattice parameters have an effect of spontaneous magnetostriction as shown in Fig. 3.22. On the other hand, when T_C is lower than room temperature ($0.06 < x < 0.08$), there is no effect of the magnetostriction. As shown in Fig. 3.22, magnetostriction of $MnAs_{0.7}Sb_{0.3}$ at $T = 80$ K is +1% for the a-axis and −0.7% for the c-axis. This corresponds to volume magnetostriction $\Delta V/V = +1.5 \times 10^{-2}$.

The experimental result that T_C takes a minimum for the composition (Fig. 3.20) is phenomenologically explained from the viewpoint of localized model similar to the model of Bean and Rodbell shown in Sect. 3.2.1.

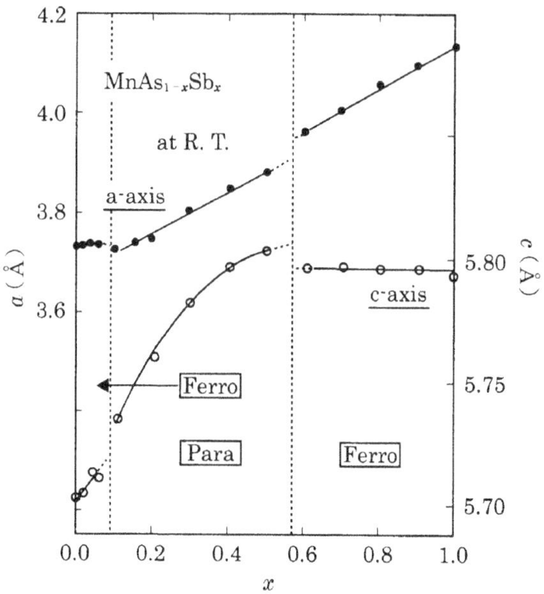

Fig. 3.21. Lattice constants of $MnAs_{1-x}Sb_x$ at room temperature [39]

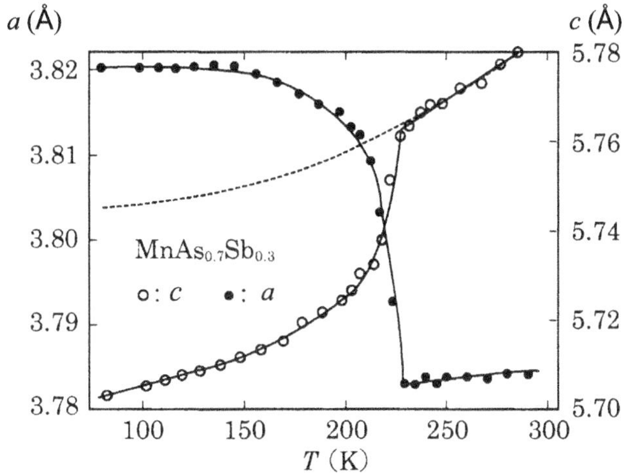

Fig. 3.22. Temperature dependence of lattice constants of $MnAs_{0.7}Sb_{0.3}$. Spontaneous magnetostriction arises below 227 K ($=T_C$) [22]

We assume that T_C is a function of exchange parameters J_c and J_a, which change linearly with lattice parameters c and a around c_0 and a_0, respectively, as follows:

$$J_a(a) = J(a_0) + A(a - a_0),$$
$$J_c(c) = J(c_0) + B(c - c_0), \tag{3.13}$$

where A and B are constants. Therefore, change of T_C due to lattice change is written as

$$dT_C(J_c, J_a) = (\partial T_C/\partial J_a)(\partial J_a/\partial a)\,da + (\partial T_C/\partial J_c)(\partial J_c/\partial c)\,dc \qquad (3.14)$$

and

$$\begin{aligned} T_C(a,c) - T_C(a_0,c_0) &= (\partial T_C/\partial J_a)A(a-a_0) + (\partial T_C/\partial J_c)B(c-c_0) \\ &= \alpha(a-a_0) + \beta(c-c_0), \end{aligned} \qquad (3.15)$$

where $(\partial T_C/\partial J_a)A$ and $(\partial T_C/\partial J_c)B$ are replaced by α and β, respectively. We insert the experimental lattice constants and T_C of MnSb for $T_C(a_0, c_0)$. By using (3.13) and the composition dependence of lattice constants shown in Fig. 3.21, we can calculate T_C of $MnAs_{1-x}Sb_x$ as plotted in Fig. 3.23. We note that for $x \geq 0.6$ (where T_C is higher than room temperature) we extrapolated the lattice constant curve (Fig. 3.21) in the region $x \leq 0.5$ so as to exclude the effect of ferromagnetism.

Results shown in Fig. 3.23 were obtained by assuming linear change of exchange parameters J_c and J_a against corresponding lattice constants. In calculation we set the coefficients $\alpha = 1360\,\mathrm{K/Å}$ and $\beta = -3000\,\mathrm{K/Å}$. With these parameters we successfully reproduced the minimum of T_C against composition of $MnAs_{1-x}Sb_x$ qualitatively. It is apparent that anomalous composition dependence of the c-axis gives rise to the composition dependence of T_C, though there is no theoretical foundation for the anomalous composition dependence of the c-axis. We also see from Fig. 3.22 that expansion of the a-axis as well as the shrink of c-axis brings about increase of T_C. This corresponds to the sign of α and β. If elastic constants along the a- and c-directions are available, it is possible to evaluate quantitatively the

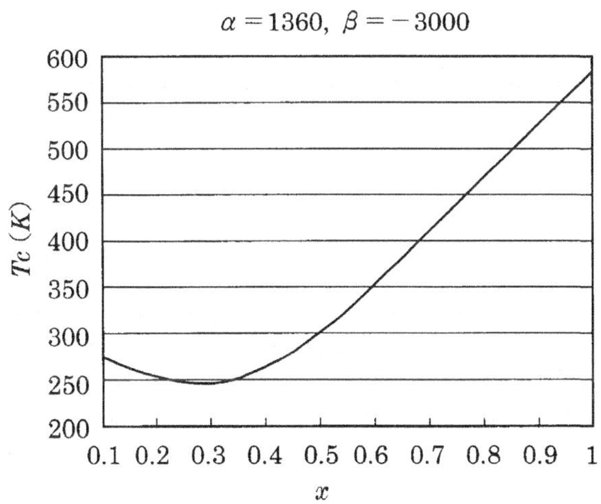

Fig. 3.23. T_C of $MnAs_{1-x}Sb_x$ calculated using J_a and J_c whose temperature change was evaluated from experimental values of the lattice constants. The minimum of T_C is caused by anomalous variation of the c-axis for the composition (Fig. 3.21 and also see text)

composition dependence of T_C as well as the temperature dependence of spontaneous magnetostriction and magnetic moment on the basis of the Bean–Rodbell theory [21]. Lattice constants of MnSb and CrSb under high pressure up to 1.8 GPa have been measured by Nagasaki [42, 43]; however, a more precise measurement using state-of-the-art equipment is expected.

Next we describe metamagnetic behavior of $MnAs_{1-x}Sb_x$. The metamagnetism of MnAs just above T_C arises from the fact that ferromagnetic–paramagnetic transition at T_C is of the first order ($\eta \geq 1$). In $MnAs_{1-x}Sb_x$, similar metamagnetic transition induced by applied magnetic field occurs only in the composition region of $x \leq 0.08$ (see Fig. 3.20). On the other hand, for $x \geq 0.1$, metamagnetic transition does not occur; however, metamagnetic-like behavior remains near $x = 0.1$, which means the parameter η in the argument in Sect. 3.2.1 is slightly smaller than 1. As an example, metamagnetic-like behavior of $MnAs_{0.7}Sb_{0.3}$ just above T_C is shown in Fig. 3.24 [22].

This field-induced metamagnetic-like transition does not accompany the structural transformation between the MnP-type and NiAs-type structures. As shown in Appendix, metamagnetic magnetization process is generally given by the expression

$$H = A_1\sigma + A_2\sigma^3 + A_3\sigma^5 + \ldots, \tag{3.16}$$

where H is a magnetic field, σ magnetization, and A_1 etc. are expansion coefficients. The form of (3.16) is common to the Bean–Rodbell model and the itinerant electron model developed by Yamada [44], though physical meaning of the coefficients is different. The experimental results shown in Fig. 3.24 are well reproduced by (3.16)

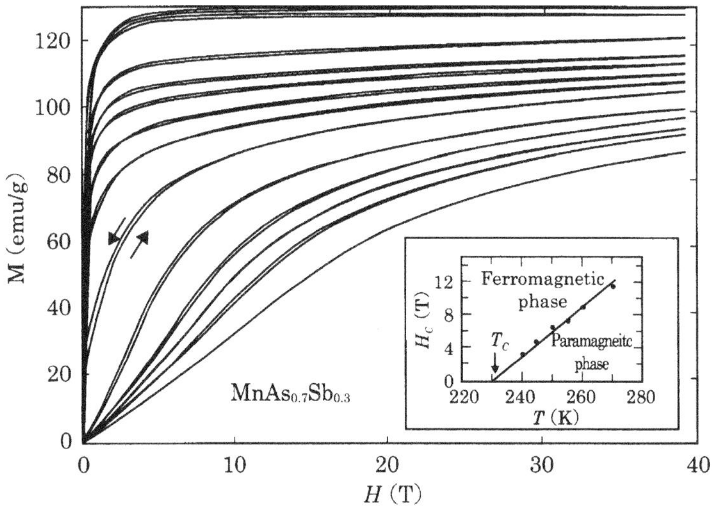

Fig. 3.24. Magnetization curves of $MnAs_{0.7}Sb_{0.3}$. The curves are measured at $T = 4.2$, 77, 180, 200, 210, 220, 225, 230, 240, 250, 255, 260, 270 K from larger magnetization. *Inset*: magnetic fields at inflection points are plotted as a function of temperature. From this figure we see that $T_C = 230$ K (Reproduced from [22])

with temperature-dependent A_1, A_2, and A_3. However, both the above two models cannot explain successfully the temperature change of A_2 [22]. We also note that there is some allowance for the experimental values of A_1, A_2, and A_3.

3.2.4 Effect of High Pressure on $MnAs_{1-x}Sb_x$

Pressure effects on T_C of MnSb and $MnAs_{0.88}Sb_{0.12}$ [45] and changes of magnetic states of $MnAs_{0.7}Sb_{0.3}$ [46] have been investigated under high pressure. T_C of MnSb decreases by 200 K (from 590 to 390 K) by applying pressure 5 GPa [47]. Under this pressure, the a- and c-axes decrease about 2% (from 4.13 to 4.03 Å) and 4% (from 5.78 to 5.55 Å), respectively [43]. This corresponds to the compressibility $K = -(1/V)(dV/dP) = 8.6 \times 10^{-11}\,\mathrm{Pa}^{-1}$ ($= 8.6 \times 10^{-12}\,\mathrm{cm}^2\,\mathrm{dyn}^{-1}$), which is an ordinary value for this kind of intermetallic compounds [42]. The elastic constant of MnSb is anisotropic and the compressibility is much larger than that of a material which has closed packed structure (e.g., four times larger than that of Cu and eight times larger than that of Ni). Unlike $MnAs_{0.7}Sb_{0.3}$ shown in Fig. 3.22, the temperature variation curve of c-axis of MnSb shows no anomaly at T_C while the a-axis shows small positive magnetostriction [48]. This fact indicates that the ferromagnetism is stabilized by the increase of a-axis. If we are based on the localized model, the positive magnetostriction means that increase of a-axis results in increase of an exchange interaction parameter J_a connecting Mn spins within the c-plane. This is consistent to the behavior of $MnAs_{1-x}Sb_x$ with $x \geq 0.6$ shown in Figs. 3.20 and 3.21, where T_C increases as a-axis increases (c-axis is almost constant) with increasing composition x. Therefore, the decrease of T_C of MnSb by application of pressure is mainly caused by the decrease of a-axis. It is apparent that there is a close correlation between magnetism and lattice parameters. To clarify the mechanism from microscopic electronic structure is a problem left for the future.

Next, we discuss pressure effect on T_C of $MnAs_{0.88}Sb_{0.12}$. Unlike MnAs, this compound shows a second order transition at T_C. As shown in Fig. 3.25, T_C decreases and temperature interval ΔT of hysteresis for magnetization increases with increasing pressure. The curve for 1.2 GPa suggests a structural change caused by the pressure from the NiAs- to the MnP-types.

$MnAs_{0.88}Sb_{0.12}$ has an Sb composition x just above the critical composition $x = 0.1$, as seen in Fig. 3.20; therefore, shows at T_C the second order transition having the characteristic near the first order transition. The hysteresis generated by applied pressure shown in Fig. 3.25 means that pressure changes the second order transition to the first order transition at T_C. (The large hysteresis shown in Fig. 3.25 indicates the first order transition though the order is unclear especially under pressures $P = 0.8$ and 1.0 GPa).

We will show that these experimental results are explained by the Bean–Rodbell model. We assume $\eta = 1$ (see (3.11) and (3.12)) as this compound locates near the border between the first order and the second order transition. Then, by using parameters $T_0 = 300$ K and $K\beta = 5.3 \times 10^{-11}$ (CGS), which are the same as those as MnAs, we can calculate by making use of (3.12) the magnetization curves plotted in Fig. 3.26.

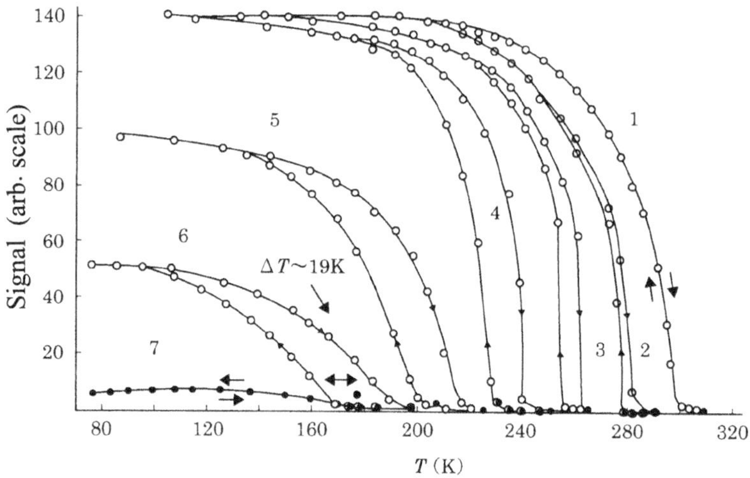

Fig. 3.25. Temperature dependence of magnetization (in arbitrary scale) of $MnAs_{0.88}Sb_{0.12}$ under high pressure. The labels 1, 2, 3, 4, 5, 6, and 7 denote the pressure 0, 0.2, 0.4, 0.6, 0.8, 1.0, and 1.2 GPa, respectively (Reproduced from [45])

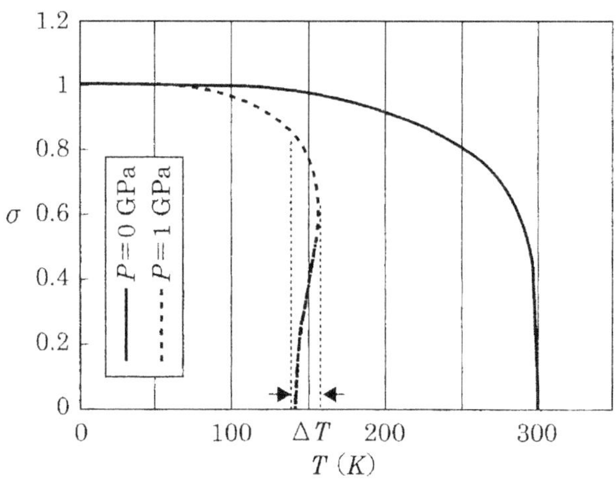

Fig. 3.26. Magnetization (σ) vs. temperature curves of $MnAs_{0.88}Sb_{0.12}$ ($\eta = 1$, $T_0 = 300$ K, $\beta K = 5.3 \times 10^{-11}$ (CGS)) under atmospheric and high pressure ($P = 1$ GPa). 15 K width of hysteresis is seen for $P = 1$ GPa

Calculated results shown in Fig. 3.26 exhibits that pressure alters order of transition from the second to the first, and the calculated values of $\Delta < 15$ K is in fairly good agreement with the experimental results of $\Delta \simeq 19$ K (Fig. 3.25), even without parameter adjustment. It is noted that the molecular field theory based on the localized picture (i.e., the Bean–Rodbell model) is successful in explaining various properties of itinerant electron system such as present compounds.

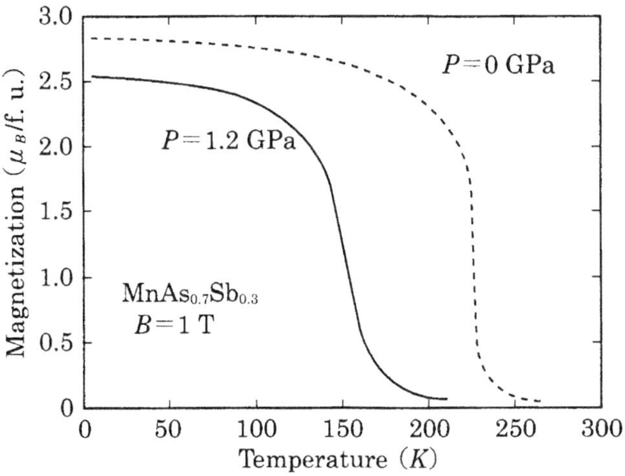

Fig. 3.27. Magnetization–temperature curves of $MnAs_{0.7}Sb_{0.3}$ under atmospheric and high pressure (1.2 GPa) (Reproduced from [46])

Next, we show pressure effect on $MnAs_{0.7}Sb_{0.3}$ in magnetic field up to 7 T and in quasi-hydrostatic pressure up to 1.2 GPa [46]. We have already noted that this material exhibits a second order transition at T_C; however, this compound shows metamagnetic-like behavior just above T_C as shown in Fig. 3.24. Figure 3.27 shows magnetization–temperature curves under pressures 0 and 1.2 GPa. We see that T_C decreases by 70 K due to 1.2 GPa pressure. (T_C of $MnAs_{0.88}Sb_{0.12}$ decreases by 105 K). It is not clear in Fig. 3.27 whether the order of transition at T_C changes by pressure from the second to the first. However, temperature dependence of magnetization under 8 and 10 GPa shown in Fig. 3.25 suggests that phase transition under 1.2 GPa shown in Fig. 3.27 is of the first order. It is possible to clarify whether the transition is of the first order or of the second by measuring the hysteresis Δ or temperature dependence of spontaneous magnetization under pressure.

Finally, we show pressure dependence of magnetization curves at $T = 4.2$ K in Fig. 3.28. It is clearly seen that spontaneous magnetization at 4.2 K decreases largely by pressure; the spontaneous moment per Mn atom $M(\mu_B)$ is found to be expressed by $M = 3.185 - 0.126P - 0.052P^2$, where P means the pressure in the unit of GPa. A relation between spontaneous magnetic moment M_s and T_C under pressure, $d\ln M_s/dP = (2/3)\, d\ln T_C/dP$, has been obtained for a weak itinerant ferromagnet by Takahashi [49]; however, this relation seems to be inapplicable to a strong ferromagnet $MnAs_{0.7}Sb_{0.3}$. Pressure dependence of spontaneous magnetization at $T = 4.2$ K is explained qualitatively by the energy band calculation for changing lattice parameters [46]. Note that there are many problems left for the future.

3.2.5 $Mn_{1-x}Cr_xAs$, $Mn_{1-x}Ti_xAs$, etc.

As MnAs is unstable against the MnP-type deformation, a mixed compound $Mn_{1-x}Cr_xAs$ (CrAs has the MnP-type structure) has the MnP-type structure in

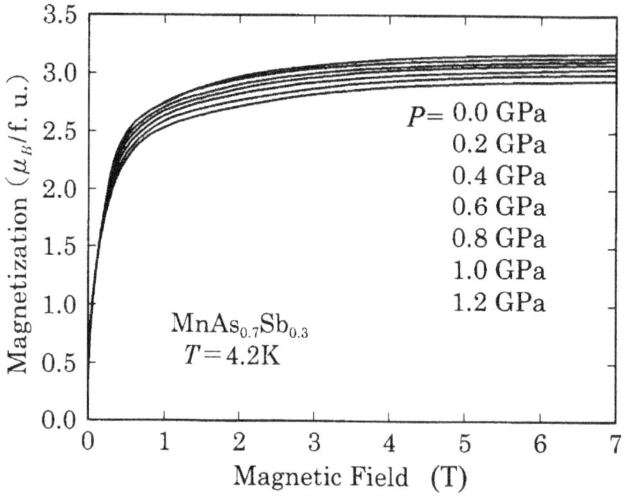

Fig. 3.28. Pressure effect on the magnetization curve of $MnAs_{0.7}Sb_{0.3}$ at $T = 4.2$ K (Reproduced from [46])

low temperature region even when x is small, and undergoes the structural transformation to the NiAs-type when temperature rises. These structural properties of $Mn_{1-x}Cr_xAs$ are similar to those of $MnAs_{1-x}P_x$ [50]; however, magnetic properties are somewhat different. Existence of two kind of $3d$ atoms in $Mn_{1-x}Cr_xAs$ makes the magnetic properties a little complex.

By substituting a nonmagnetic atom Ti for Mn of MnAs and MnSb, density of magnetic atom in the compound decreases and the NiAs-type structure is stabilized, except the vicinity of TiAs (TiSb). T_C of these mixed compounds shows rapid decrease with Ti composition, approaching 0 K for $x = 1$. $Mn_{1-x}Ti_xSb$ becomes spin glass in low temperature for the compositions around $x = 0.8$ [51]. Subtracting Pauli paramagnetic contribution ($\chi_p = 1 \times 10^{-6}$ emu g^{-1}), susceptibility of $Mn_{1-x}Ti_xAs$ obeys the Curie–Weiss law [14]. Form this Curie–Weiss type susceptibility, the paramagnetic Curie temperature Θ_p and the effective Bohr magneton per Mn atom, $\mu_{\mathrm{eff}} = g\mu_B\sqrt{S(S+1)}$, were evaluated. By setting the g-factor 2 in μ_{eff}, we can calculate $gS\mu_B$ (S : spin) $= P_A$ (atomic moment). Figure 3.29 illustrates Θ_p, T_C, and P_A as a function of Ti composition. $Mn_{1-x}Ti_xSb$ also shows the similar behavior [14]. It is notable that P_A is almost independent on Ti composition. There has been report on temperature dependence of lattice constants and pressure effect on T_C of $Mn_{1-x}Ti_xAs$ ($0 \le x \le 0.9$) [21]. An itinerant electron theory is desired to explain the properties shown above.

3.2.6 Magnetic Refrigeration Using MnAs and the Related Compounds

If an applied magnetic field largely reduces entropy of a material, this material can be a candidate for a refrigerant of magnetic refrigeration. Enhancement of

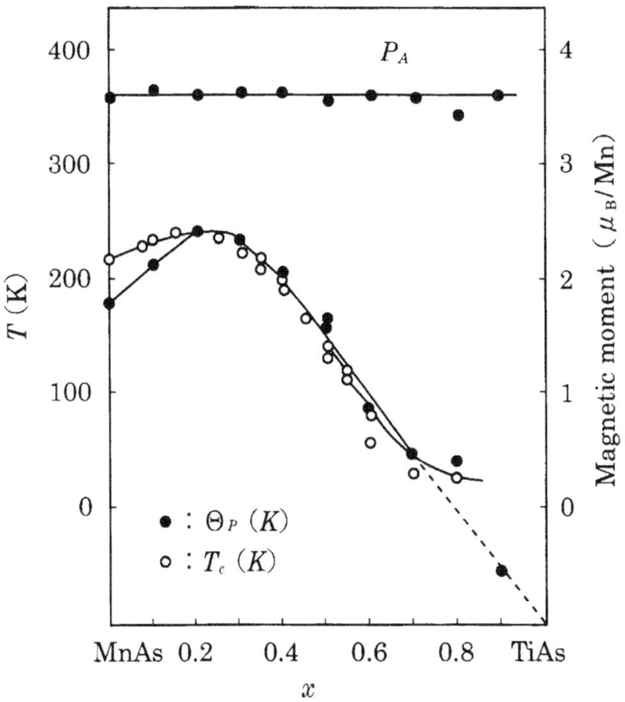

Fig. 3.29. T_C, θ_p, and $P_A (= gS\mu_B/\text{Mn}$, see text) of $Mn_{1-x}Ti_xAs$. $Mn_{1-x}Ti_xSb$ shows similar behavior (Reproduced from [14])

magnetization due to applied magnetic field means increase of magnetic order or reduction of magnetic entropy. For a ferromagnet, large increase of magnetization due to applied field occurs in the temperature region around T_C; just above and below T_C for the second order transition and only just above T_C for the first order transition (cf. Figs. 3.4 and 3.6). A ferromagnet that shows a first order transition (e.g., MnAs) shows metamagnetic transition just above T_C. Metamagnetic transition field H_C increases with rising temperature (see Fig. 3.7). In Fig. 3.30, magnetization curves above and below T_C are show schematically for a ferromagnet with the second or the first order transitions at T_C.

If we measure or calculate magnetization curves at various temperatures, it is possible to evaluate corresponding magnetic entropy using the expansion formula of σ (3.3).[3] In an adiabatic condition where total entropy is constant, reduction of

[3] Temperature dependence of decrease of magnetic entropy ΔS_M by application of external field H is calculated from difference of specific heats in $H = 0$ and $H = H$. ΔS_M is also calculated by integrating the Maxwell relation $(\partial S/\partial H)_T = (\partial M/\partial T)_H$ of thermodynamics as

$$\Delta S_M = \int_0^H (\partial M/\partial T)_H \, \mathrm{d}H,$$

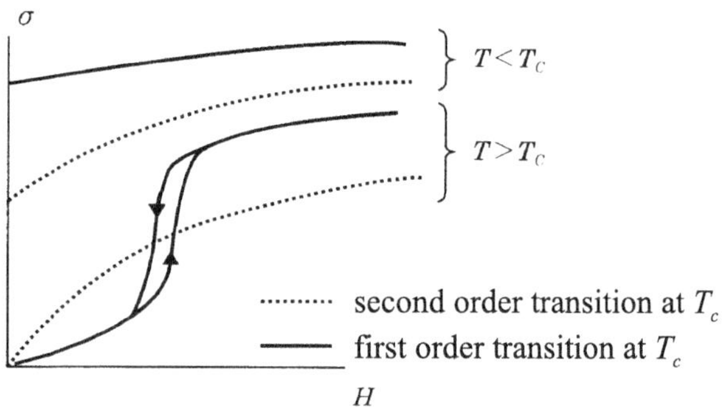

Fig. 3.30. Schematic magnetization curves above and below T_C

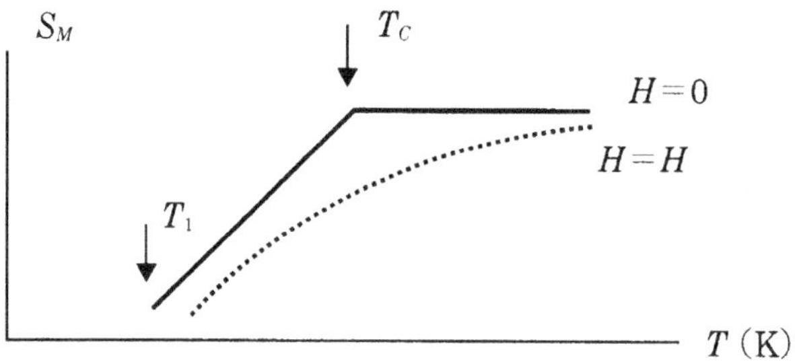

Fig. 3.31. Temperature dependence of magnetic entropy (a schematic figure for the case of a second order transition). See text for T_1

magnetic entropy due to applied field enhances entropy of lattice vibration, bringing about increase of temperature by ΔT. This is called magnetocaloric effect. A material that has large ΔT is suitable for magnetic refrigerant.

We explain briefly how ΔT arises below [52]. Figure 3.31 shows schematic temperature dependence of magnetic entropy S_M with and without magnetic field.

Total entropy S_T of a magnetic material is $S_M + S_L$ (S_L: lattice entropy). Assuming that the lattice specific heat C_L is constant in the range $T_1 \leq T$ (T_1 is a temperature just below T_C), S_L is calculated as

$$S_L(T) - S_L(T_1) = \int_{T_1}^{T} C_L \, dT/T = C_L \ln(T/T_1). \tag{3.17}$$

This equation means that the lattice entropy is a monotonically increasing function of temperature. Taking account of temperature dependence of S_M in Fig. 3.31, we can express S_T schematically as shown in Fig. 3.32.

where M is magnetization under a magnetic field. We need to measure magnetization in various magnetic fields to obtain ΔS_M. By integrating $\partial M/\partial T$ with respect to H at respective temperatures, we can calculate temperature variation of ΔS_M. We note that M in this argument means magnetization in a state in which all the atomic moments align in the direction of H, that is, a state without effect of anisotropy and magnetic domain.

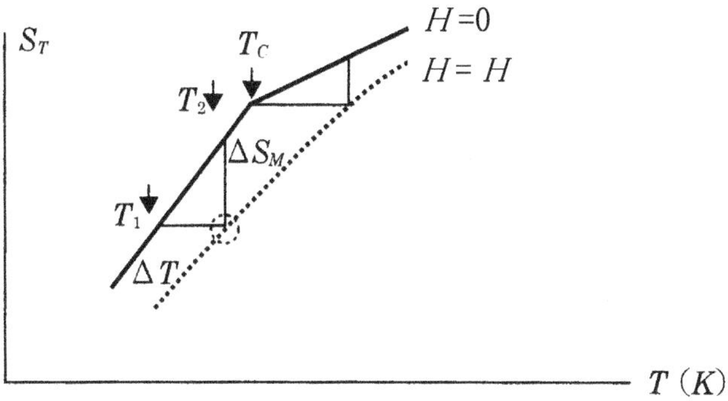

Fig. 3.32. Schematic curves of total entropy S_T vs. temperature T in $H = 0$ and $H = H$

If we apply H adiabatically (i.e., $S_{\rm T}$ is unchanged) to a magnetic material at temperature $T = T_1$ $(H = 0)$, temperature rises up to $T = T_2$ by ΔT. The resulting state at $T = T_2$ $(H = H)$ is equivalent to a state in which H is isothermally applied to a material at $T = T_2$ $(H = 0)$. In the former adiabatic process, entropy of spins decreases (because spins align along the applied field) and lattice entropy increases, and then temperature of the material increases. On the other hand, dissipation of heat keeps the temperature unchanged in the latter isothermal process.

The process of magnetic refrigeration is explained next. By applying H, the state $(T_2, H = 0)$, which is taken as an initial state here, changes to the state $(T_2, H = H)$. In this process, generated heat is released to outside. Adiabatic removal of magnetic field brings the state (T_2, H) to $(T_1, H = 0)$. In this process, the temperature of the material lowers from T_2 to T_1, namely the material absorbs heat from outside (i.e., air around). Repetition of the similar process is the magnetic refrigeration. The magnetic material must show large magnetocaloric effect within a finite temperature range because the temperature of refrigerant decreases during the refrigeration process. A magnetic material that has large ΔT in a wide temperature range is favorable. From Fig. 3.32 we see that $\mathrm{d}S_{\rm T}/\mathrm{d}T = \mathrm{d}S_{\rm L}/\mathrm{d}T \simeq \Delta S_{\rm M}/\Delta T$ for $T \geq T_{\rm C}$. Since $\mathrm{d}S_{\rm L}/\mathrm{d}T = C_{\rm L}/T$ as seen in (3.17), we have

$$\Delta T \simeq T\Delta S_{\rm M}/C_{\rm L} \simeq T_{\rm C}\Delta S_{\rm M}/C_{\rm L}. \tag{3.18}$$

Therefore, ΔT is proportional to reduction of magnetic entropy due to applied field. For $T \leq T_{\rm C} - \Delta T$, $\mathrm{d}S_{\rm T}/\mathrm{d}T = \mathrm{d}S_{\rm L}/\mathrm{d}T + \mathrm{d}S_{\rm M}/\mathrm{d}T = C_{\rm L}/T + C_{\rm M}/T$ is approximately satisfied and then

$$\Delta T \simeq T_{\rm C}\Delta S_{\rm M}/(C_{\rm L} + C_{\rm M}(H = 0)), \tag{3.19}$$

where $C_{\rm M}$ $(H = 0)$ is the magnetic specific heat in $H = 0$. From (3.18) and (3.19) we see that ΔT for $T < T_{\rm C}$ is expected to be smaller than ΔT for $T > T_{\rm C}$ because of the difference between denominators of (3.18) and (3.19). Wada et al. investigated MnAs and $\mathrm{MnAs_{1-x}Sb_x}$ as magnetic refrigerant [53–55]. We show some of their data in Fig. 3.33.

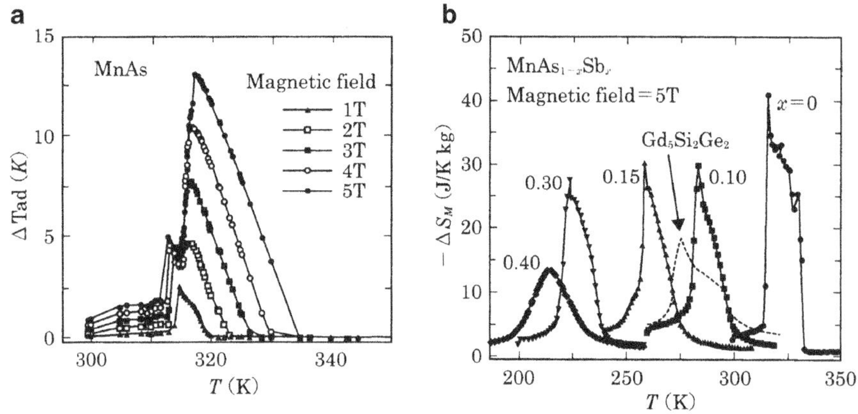

Fig. 3.33. (**a**) Magnetocaloric effect ΔT of MnAs evaluated indirectly from temperature variations of total entropy S_T and magnetization. (**b**) Decrease of magnetic entropy $-\Delta S_M$ of $MnAs_{1-x}Sb_x$ measured by applying magnetic field 5 T (Reproduced from [53])

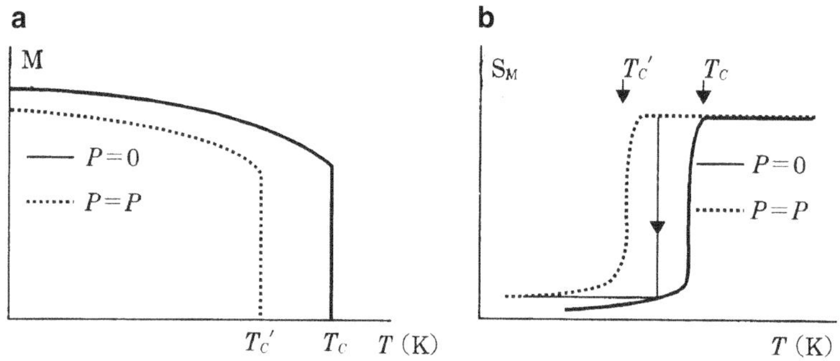

Fig. 3.34. Schematic figure to show the mechanism of refrigeration by pressure for $MnAs_{1-x}Sb_x$ with small x. This figure is for a first order transition at T_C

In Fig. 3.33b, ΔS_M at $H = 5$ T was evaluated from temperature change of magnetization in various magnetic fields, whereas in (a), ΔS_M was determined indirectly from the data shown in (b) and temperature dependence of entropy in zero field [55]. Method of direct measurement of the magnetocaloric effect and experimental data for Gd metal have been reported together with analysis based on the molecular field theory in [56].

We have discussed in Sect. 3.2.4 that T_C of $MnAs_{1-x}Sb_x$ with small Sb content decreases remarkably by application of pressure. It may be possible to apply the compounds for refrigeration using pressure instead of magnetic field for the refrigeration process. As seen from Fig. 3.34b, removal of pressure isothermally reduces magnetic entropy. This is equivalent to application of magnetic field isothermally. It is interesting to clarify which is more practical.

3.3 MnSb and MnBi

We have already shown basic properties of these ferromagnetic compounds in Sects. 1.4 and 2.1. As MnSb has already been mentioned in Sect. 3.2.4, we explain MnBi in this section. As shown in Table 1.2 and Fig. 3.11, MnBi takes the NiAs-type structure with lattice constants larger than MnSb, and magnetic moment per Mn atom is $3.8\mu_B$. MnBi shows a first order transition with temperature hysteresis 15 K at $T_C = 628$ K [57]. Figure 3.35 shows temperature dependence of spontaneous magnetic moment, where data not only for low temperature phase (LTP) MnBi but also for $Mn_{1.08}Bi$ (quenched high temperature phase: QHTP) are plotted. The LTP shows a first order transition at T_C, which is accompanied with crystallographic transition to the nonstoichiometric $Mn_{1.08}Bi$. The QHTP has smaller moment and lower T_C than those of the LTP MnBi. These properties are similar to those of $Mn_{1+x}Sb$; however, x is limited to 0.08 for $Mn_{1+x}Bi$.

The QHTP MnBi separates into Mn and Mn-Bi liquid at the peritectic temperature of 446°C. As MnBi has large magnetic anisotropy with magnetic moment parallel to the c-axis (anisotropy constant $K \simeq 2 \times 10^7$ erg cm^{-3} $= 2 \times 10^6$ J m^{-3}) [57], there were attempts for applications to permanent magnet [58]. However, this material is not suitable for a magnet because of the relatively small magnetization and

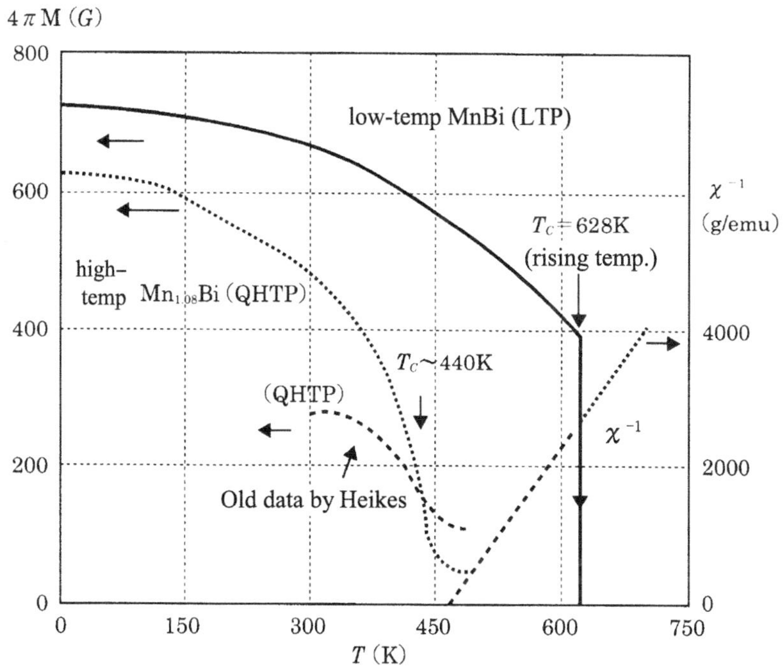

Fig. 3.35. Spontaneous magnetization $4\pi M$ and inverse susceptibility χ^{-1} of MnBi ($T \leq T_C$) and $Mn_{1.08}Bi$ (high temperature phase). Old data are also shown by the dashed curve for comparison

difficulty in preparation of compound. This material may be utilized for other type of application such as magnetic memory. There has been no theoretical study to explain the peculiar properties of MnBi.

3.4 CrAs and Related Compounds

3.4.1 Anomalous Magnetic Transition of CrAs

In contrast with the ferromagnetic Mn-pnictides with MnX-type, the compounds with CrX (X = As or Sb) type take generally antiferromagnetic ordering. As magnetic properties of CrAs are especially interesting, we describe mainly CrAs and its related compounds.

As explained in Sect. 2.1, CrAs takes the double helical magnetic ordering below $T_N = 272$ K [59], and undergoes a first order transition at T_N to paramagnetic state in high temperature region above T_N. We show susceptibility of CrAs and $CrAs_{0.9}P_{0.1}$ in Fig. 3.36 (Ido, unpublished). Similar data for $CrAs_{1-x}P_x$ have already been reported by Selte [60].

As seen in Fig. 3.36, susceptibility of CrAs increases with increasing temperature without anomaly at T_N, suggesting strong itinerant character. It is known that CrAs with the MnP-type crystal structure takes the NiAs-type structure in the temperature region higher than $T_t = 1100$ K, and the magnetic susceptibility obeys the Curie–Weiss law in this temperature range, which is not included in Fig. 3.36 [50]. For $CrAs_{1-x}Sb_x$ ($x \leq 0.6$), the Curie–Weiss law in the region of $T \geq T_t$ is clearly observed as T_t decreases due to substitution of Sb for As [61]. For example, $CrAs_{0.9}Sb_{0.1}$ has transition temperature $T_t = 780$ K and shows the Curie–Weiss behavior of susceptibility in $T \geq 780$ K. From the Curie–Weiss law, moment of Cr in $CrAs_{0.9}Sb_{0.1}$ is evaluated from the effective Bohr magneton to be $2.8\mu_B$, which is equal to the

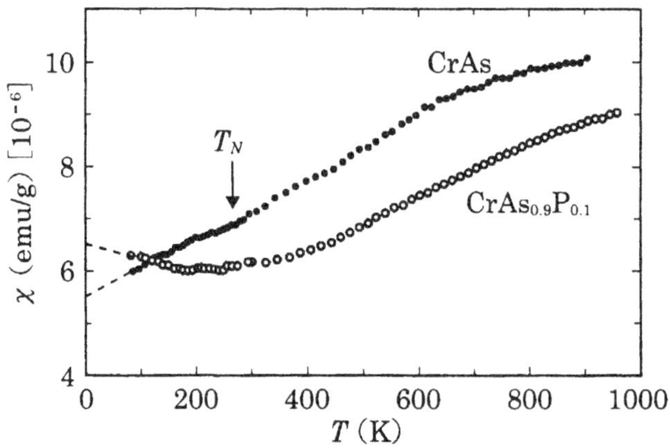

Fig. 3.36. Susceptibility of CrAs and $CrAs_{0.9}P_{0.1}$. The magnetic order occurs in CrAs below T_N, while no magnetic order occurs in $CrAs_{0.9}P_{0.1}$ down to 0 K

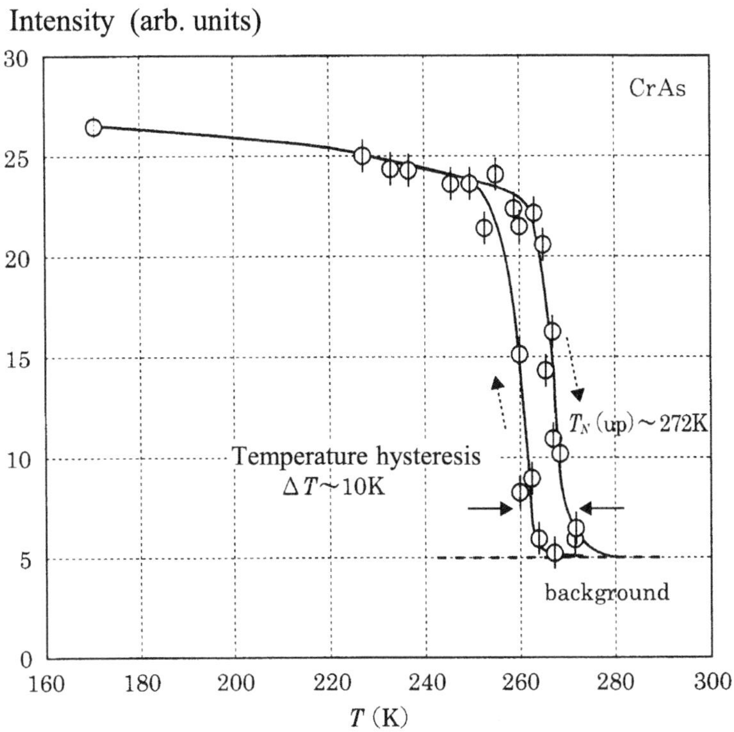

Fig. 3.37. Temperature change of integrated intensity of (002) satellite of CrAs (Reproduced from [59])

moment of Cr in CrSb determined by neutron diffraction measurements (see also Sect. 3.5). It is interesting that susceptibility of CrAs in the NiAs-type phase obeys approximately the Curie–Weiss law expected for Cr^{3+} (i.e., $3\mu_B$). This phenomenon will be shown and discussed later in Fig. 3.42. T_N of CrAs was determined by neutron diffraction measurements (Fig. 3.37) [59, 62]. Magnetic phase transition at T_N is of the first order accompanied with discontinuous change of lattice constants. We will discuss in detail later the magnetovolume effect of CrAs.

It has been found from specific heat measurements that latent heat accompanied with magnetic phase transition is 5.3×10^2 J mol^{-1} (=0.98 cal g^{-1}) [63,64]. Temperature dependence of electric resistivity is similar to that of a gap-type antiferromagnet such as Cr and $Cr_{1-x}Mn_x$ ($x \leq 0.25$), whose resistivity increases first with decreasing temperature below T_N and decreases next on further cooling [63]. It is difficult to measure resistivity of CrAs accurately because of cracks arising due to large volume change at T_N. This is similar to the situation in MnAs.

In Fig. 3.38, magnetic moment per Cr atom for some Cr pnictides is plotted as a function of length of the a-axis. This figure is similar to Fig. 3.11, where Mn-moment is plotted against a-axis length for the Mn-pnictides. We note again that the a-axis

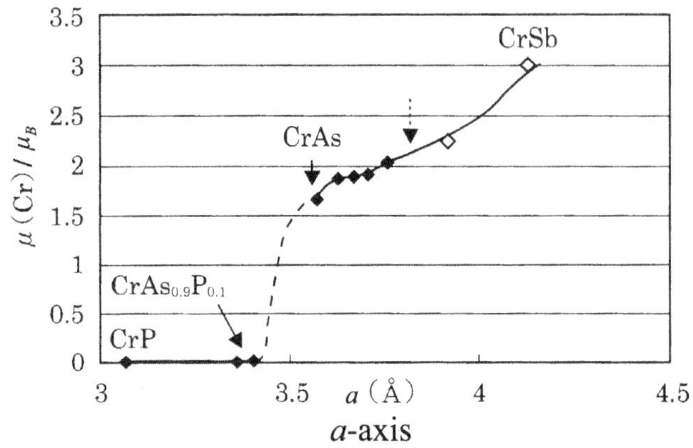

Fig. 3.38. Relation between moment per a Cr atom and the a-axis at 4.2 K. *Filled* and *open* symbols denote the MnP-type and NiAs-type structures, respectively. This figure has been drawn by making use of the data in [65] and [66, 67].

for the orthorhombic MnP-type structure in Fig. 3.38 corresponds to the a-axis of the hexagonal NiAs-type structure. The reason why we take only a-axis as a parameter for μ(Cr) in Fig. 3.38 is that the c-axis variation from 5.45 Å of CrSb to 5.36 Å of CrP is negligibly small compared with the large a-axis variation from 4.13 Å of CrSb to 3.11 Å of CrP.

Figure 3.38 shows apparently that magnetism of CrAs is very sensitive to the change of a-axis, which will be discussed in detail in the next section. It may need to point out that there is little change of μ(Cr) around $a = 3.75$ Å where the structural change occurs. This may mean that the distortion to the MnP-type structure does not affect so much. This is sharply different from $MnAs_{0.9}P_{0.1}$, whose moment exhibits large change accompanied with the structural change as shown in Fig. 3.14.

3.4.2 Critical Lattice Constant of $CrAs_{1-x}P_x$ and $Cr_{1-x}M_xAs$ (M = Mn, Ni, etc.)

We see the anomalous magnetic properties of CrAs more clearly in the behavior of mixed compounds. A typical example is a CrAs-based compound in which small fraction of As is replaced by P. Both electron number and crystal structure remain unchanged by the P substitution for As. According to [60], the double helical ordering with the first order transition at $T_N \simeq 250$ K disappears for $CrAs_{1-x}P_x$ with $x \geq 0.03$ [60], which is also roughly expected from Fig. 3.38. This is a very interesting phenomenon whose mechanism has not been clarified yet.

We show and discuss experimental data of magnetovolume effect of $CrAs_{1-x}P_x$ in the following part of this section. Figure 3.39 shows temperature change of lattice constants of some of the mixed compounds derived from CrAs [68]. As we noted in Sect. 1.3, the A-, B-, and C-axes here correspond to the c-, a-, and b-axes of the hexagonal NiAs-type structure, respectively.

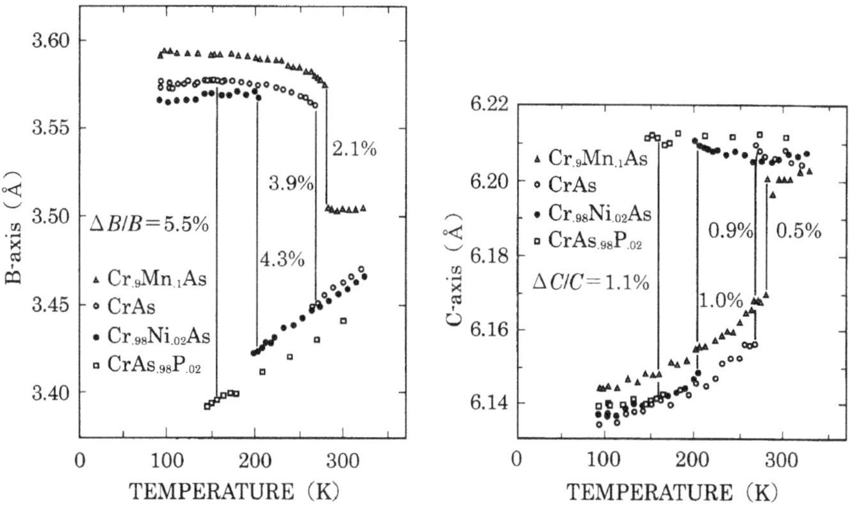

Fig. 3.39. Temperature change of lattice constants of CrAs and related compounds. Note that the *B*- and *C*-axes correspond to the *a* and *b*-axes of the hexagonal NiAs-type structure, respectively (Reproduced from [68])

In Fig. 3.39 data for the $A(c)$-axis, where (c) means the A-axis corresponds to the c-axis, are not plotted because the discontinuous change of $A(c)$-axis at T_N is much smaller than those of $B(a)$- and $C(b)$-axes. For example, in the case of $CrAs_{0.98}P_{0.02}$, the discontinuous change of the B- and C-axes are 5.5% and 1.1%, respectively, while that of the $A(c)$-axis is only 0.4%. Therefore, as generally seen in Fig. 3.39, magnetovolume effect takes place mainly in the $B(a)$-direction. It is noted that the change of the B-axis is exceptionally large. We may regard the change in the $C(b)$-direction at T_N, which has an opposite sign to that of $B(a)$-axis, is only induced by the large change of the $B(a)$-axis. To obtain the normal thermal expansion curve of the $B(a)$-axis, temperature dependence of the $B(a)$-axis of $CrAs_{0.9}P_{0.1}$, which shows no magnetic ordering at $T = 0$ K, was measured for the temperature from 80 to 330 K. As shown by the dotted curves in Fig. 3.40, the part of the curve below 80 K was extrapolated on the basis of the specific heat (H. Ido and T. Kamimura, unpublished) and the Grüneisen relation [69], which says that specific heat is proportional to a thermal expansion coefficient. By applying the normal thermal expansion curve presumed for $CrAs_{0.9}P_{0.1}$, magnetovolume effect at $T = 0$ K has been estimated for CrAs and $CrAs_{0.98}P_{0.02}$ to be 4.6% (0.16 Å) and 5.6% (0.19 Å), respectively, as shown in Fig. 3.40. These values are extremely large. As no magnetic ordering occurs in $CrAs_{0.9}P_{0.1}$, this compound shows no magnetovolume effect at $T = 0$ K. We see from Fig. 3.39 that magnetic ordering of CrAs and $CrAs_{0.98}P_{0.02}$ is stabilized by the large lattice distortion in the $B(a)$-direction. We can also say that increase of elastic energy due to magnetostriction is larger than decrease of magnetic energy in $CrAs_{0.9}P_{0.1}$; therefore, this compound has no magnetic ordering. Therefore, in CrAs and the related compounds there seems to be a close relation between the occurrence

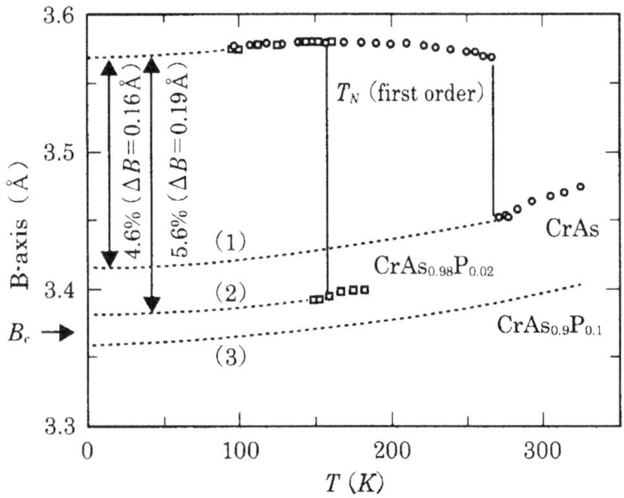

Fig. 3.40. Spontaneous magnetostriction of $CrAs_{1-x}P_x$ and critical B-axis B_C (Reproduced from [68])

of magnetic order and the length of $B(a)$-axis. As indicated in Fig. 3.40, the critical length of $B(a)$-axis, B_C, is about 3.37 Å, namely magnetic ordering occurs in the compounds with B larger than B_C, and no magnetic ordering occurs if $B < B_C$. CrAs with B larger than B_C has moment $1.7\mu_B$ and $T_N = 250$ K. Figure 3.40 indicates that magnetic ordering of CrAs is expected to disappear by more than 5.6% compression of the $B(a)$-axis. Figure 3.40 also suggests that, if spontaneous magnetostriction were fixed at the value just below the first order transition point T_N, the Néel temperature would be more than 350 K.

As seen from Fig. 3.39, T_N (a first order transition) increases due to Mn substitution for Cr of CrAs. In Fig. 3.41a, T_N determined by thermal expansion curve is plotted against composition of various substitution elements. We see that T_N decreases (increases) when the $B(a)$-axis shrinks (increases) due to the substitution. On the other hand, in Fig. 3.41b, T_N is plotted not for composition but for the length of $B(a)$-axis just above T_N (=250) K, which is almost close to the B-axis length at $T = 0$ K in the absence of magnetostriction.

As seen from Fig. 3.40, the important point we see from Fig. 3.41b is that there is a critical length of the B-axis $B_C = 3.38$ Å, which is independent of the kind of substituting atom. To construct a magnetic phase diagram, the crystallographic transition temperatures T_t between the MnP-type and the NiAs-type, which are taken from Fig. 3.42, are also shown in Fig. 3.41b.

In Fig. 3.41b, the labels "Pauli para" and "Curie–Weiss type para" indicate the regions that magnetic susceptibility behaves as those in Fig. 3.36 and as those above T_t in Fig. 3.42, respectively [61]. In Fig. 3.42, we show moment of a Cr atom μ_{Cr} (evaluated from the Curie–Weiss law of susceptibility), Θ_P, etc. As $CrAs_{0.9}Sb_{0.1}$ has higher T_N and lower T_t compared with CrAs (see Fig. 3.36 and [50]), the Curie–Weiss-type

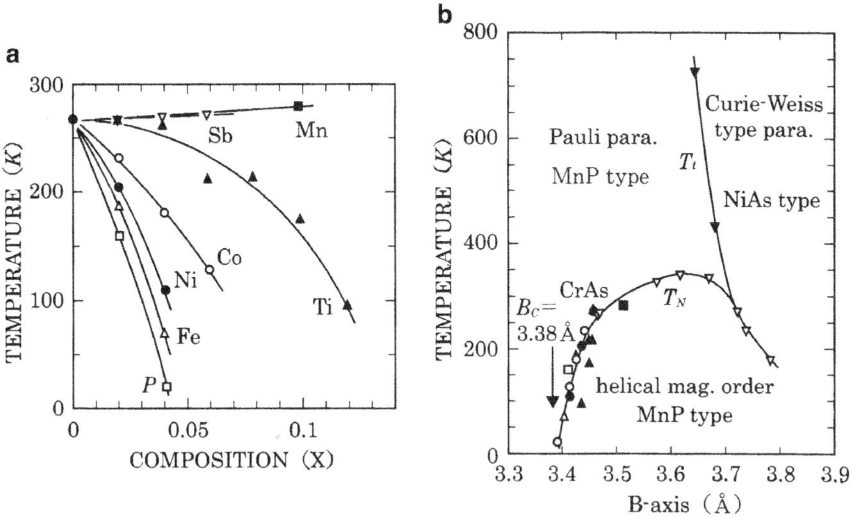

Fig. 3.41. (**a**) Composition dependence of T_N when Cr or As of CrAs is replaced by other elements, which are denoted by the following symbols: *filled triangle* (Ti); *filled square* (Mn); *open triangle* (Fe); *open circle* (Co); *filled circle* (Ni); *open square* (P); *open inverted triangle*, *filled inverted triangle* (Sb). (**b**) T_N of these compounds plotted as a function of length of the B-axis just above T_N. Data in Fig. 3.42 has also been used to make this figure. T_t is transition temperature of the structural transformation. See also the text (Reproduced from ref. [68])

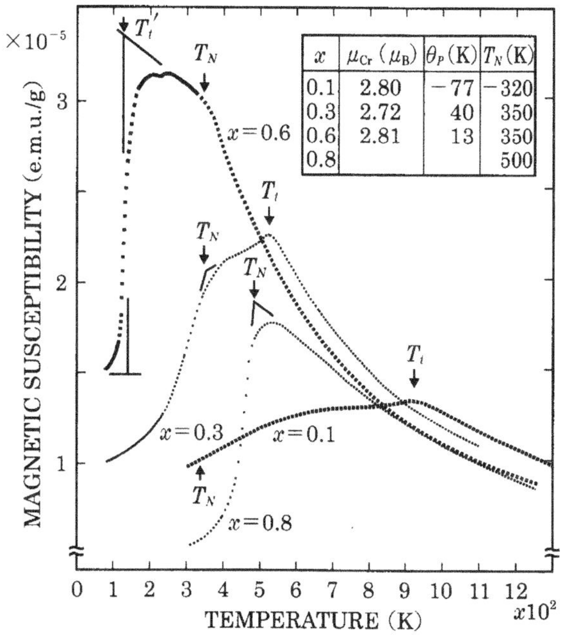

x	μ_{Cr} (μ_B)	θ_P(K)	T_N(K)
0.1	2.80	−77	−320
0.3	2.72	40	350
0.6	2.81	13	350
0.8			500

Fig. 3.42. Susceptibility of $CrAs_{1-x}Sb_x$ ($x \leq 0.6$). T_N and T_t' are the Néel temperature and temperature of the first order structural transformation (see also Fig. 3.43) (Reproduced from [61])

behavior of susceptibility appears more clearly especially for the compound with $x = 0.3$. We will discuss again the results shown in Fig. 3.42 in Sect. 3.4.4.

3.4.3 Phenomenological Theory for the First Order Transition of CrAs

The phenomenological theory for the first order transition of MnAs (see Sect. 3.2.1 and Appendix) is applicable to CrAs. We note that physical meaning of the terms of free energy expansion is different. When magnetization of space variation is written as $M(r) = M_q \exp(iqr)$ ($q = 2\pi/2.67b$ along the b-axis in the case of CrAs. See Fig. 2.1), we may assume molecular field AM_q within the molecular field theory. Then we have magnetic energy $E_m = -(1/2)AM_q^2$. By setting an assumption that the molecular field coefficient A depends linearly on crystal volume as $A = A_0[1 + \beta(V - V_0)/V_0)]$, we have an expression

$$E_m = -(1/2)A_0[1 + \beta(V - V_0)/V_0)]\sigma^2, \tag{3.20}$$

where we wrote $M_q = \sigma$, which corresponds to the relative magnetization in the Bean and Rodbell theory for MnAs (see Sect. 3.2.1). As mentioned in Sect. 3.2.1, elastic energy due to lattice distortion is written as

$$E_{el} = (1/2K)[(V - V_0)/(V_0)]^2 \tag{3.21}$$

and magnetic entropy per unit mass of the crystal is given by

$$S(\sigma) = Nk_B[S(0) + a\sigma^2 + b\sigma^4 + c\sigma^6 + \ldots]. \tag{3.22}$$

By setting derivative of $E = E_m + E_{el}$ with respect to V zero, we have a relation

$$(V - V_0)/V_0 = (K\beta A_0/2)\sigma^2. \tag{3.23}$$

Inserting (3.23) into E $(= E_m + E_{el})$, we have an expression of free energy G, which is minimized with respect to V as

$$G/Nk_BT_0 = -tS(0) - a(t-1)\sigma^2 - b(t-\eta)\sigma^4 - ct\sigma^6 + \ldots, \tag{3.24}$$

where T_0, t, and η are expressed by the following equations (3.25) and (3.26), respectively,

$$-(A_0/2aNk_B) = T_0 \quad \text{and} \quad T/T_0 = t \tag{3.25}$$

and

$$\eta = -(a^2/2Nk_Bb)K\beta^2T_0 \ (> 0). \tag{3.26}$$

We see that (3.26) is the same as (A8) in Appendix; therefore, a first order transition occurs when $\eta > 1$, namely if the parameter expressing volume dependence of magnetic energy of CrAs is large enough for $\eta > 1$ to hold, the magnetization σ $(= M_q)$ disappears discontinuously at T_N. As (3.23) shows that the spontaneous magnetization is proportional to β, the experimental results shown in Fig. 3.39 clearly show that magnetic energy is very sensitive to crystal volume (mainly to the $B(a)$-axis). Results of neutron diffraction measurements revealed that σ of the double helical CrAs at T_N is about 0.7, which is close to the value of MnAs. Spontaneous magnetostriction $(V - V_0)/V_0$ at T_N is 0.022 for CrAs [68] and 0.021 for MnAs. If we express (3.23) as $(V - V_0)/V_0 = A\sigma^2$, the coefficient A of CrAs is very close to A of MnAs; however, there is contribution of structural transformation to the magnetostriction at T_C in the case of MnAs. Therefore, we may regard the value of A for CrAs as twice as that of MnAs. The magnetism accompanied with the first order transition at T_N of CrAs, which is closely correlated to the critical length B_C of the B-axis, is a very interesting problem, and is expected to be clarified in the future in the view point of itinerant electron picture.

3.4.4 $CrAs_{1-x}Sb_x$

As noted in Sect. 2.1, CrSb takes the NiAs-type structure. In contrast to $CrAs_{1-x}P_x$, substitution of Sb for As in CrAs reduces T_t (i.e., transition temperature between the MnP-type and the NiAs-type structure). See also Fig. 3.42. Figure 3.43 shows a magnetic and crystallographic phase diagram of the CrAs–CrSb system constructed from temperature dependence of lattice constants obtained by X-ray diffraction measurements [61].

The rise of T_N in the region near CrAs seen in Fig. 3.43 is probably arising from the competition between magnetic energy and elastic energy due to magnetostriction; decrease of magnetic energy and increase of elastic energy is caused by the spontaneous magnetization in CrAs; however, increase of lattice constants due to the Sb-substitution reduces the increase of the elastic energy, resulting in stabilization of magnetic ordering and higher T_N. By this mechanism the experimental results in the region near CrAs is explained. A first order transition occurs when there is competition between the increase of elastic energy and the decrease of magnetic energy. Therefore, if there is no such competition, order of transition becomes the second, which is realized in the Sb-composition x between 0.1 and 0.4 as seen in Fig. 3.43. T_N seems to take a minimum around $x \simeq 0.5$. This behavior is similar to that in the MnSb–MnAs system; however, may have more complex origin.

Next, we explain T_t' for $x = 0.6$ as an example in Fig. 3.43. On heating the lattice constants shows at T_t', a large jump as $\Delta c = 0.15$ Å $(\Delta c/c = 0.026)$, and $\Delta a = -0.07$ Å $(\Delta a/a = -0.018)$. We note that we defined T_t' as a middle point of temperature region about 100 K wide where a high-temperature phase and a low-temperature phase coexist. Corresponding to the jump of the lattice constants at T_t', susceptibility shows steep increase at T_t' as shown in Fig. 3.42. The first order transition at T_t' is quite interesting but left unsolved.

We pay our attention again to the temperature dependence of susceptibility of $CrAs_{0.9}Sb_{0.1}$ shown in Fig. 3.42 in Sect. 3.4.2. Behavior of susceptibility in the temperature region below T_t is very similar to that of CrAs (see Fig. 3.36). On the other

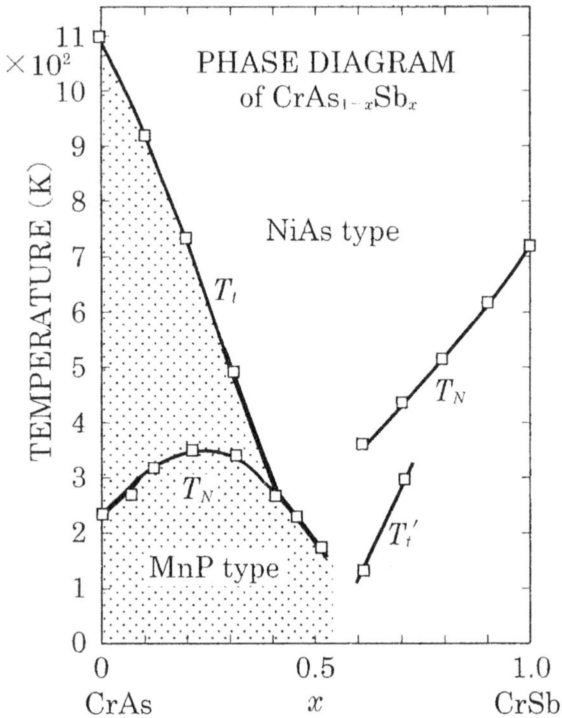

Fig. 3.43. A magnetic and crystallographic phase diagram of CrAs–CrSb system. *Thick lines* indicate that the first order transition takes place. T_t' corresponds to a transition that accompanies discrete change of the c- and a-axes; however, detail of this transition has not been revealed yet (Reproduced from [61])

hand, susceptibility for $T \geq T_t$ obeys the Curie–Weiss law with Cr moment $2.8\mu_B$ (estimated using $g = 2$) and paramagnetic Curie temperature $\theta_P = -77$ K. These values agree well with the values of CrSb [70]. It is noted that $CrAs_{0.9}Sb_{0.1}$ in the temperature region of $T \leq T_t$, where the crystal structure is with MnP-type, is similar to CrAs, while it is similar to CrSb in $T \geq T_t$, where the crystal structure is of the NiAs-type.

Finally, in Fig. 3.44, we show magnetic moment per Cr atom in the CrAs–CrSb system determined by Kallel et al. from their neutron diffraction measurements at $T = 4.2$ K [65]. The discrepancy around $x = 0.6$ is caused by the differences in magnetic order and crystal structure; on the CrSb side, the compounds have an antiferromagnetic order similar to that of CrSb with the NiAs-type structure, while on the CrAs side the compounds take the double helical ordering similar to that of CrAs with the MnP-type structure. However, change of magnetic moment at around $x = 0.6$ is not large, which is in contrast to the case of $MnAs_{1-x}P_x$, where large difference of moment between the NiAs-phase and the MnP-phase occurs (see Fig. 3.14).

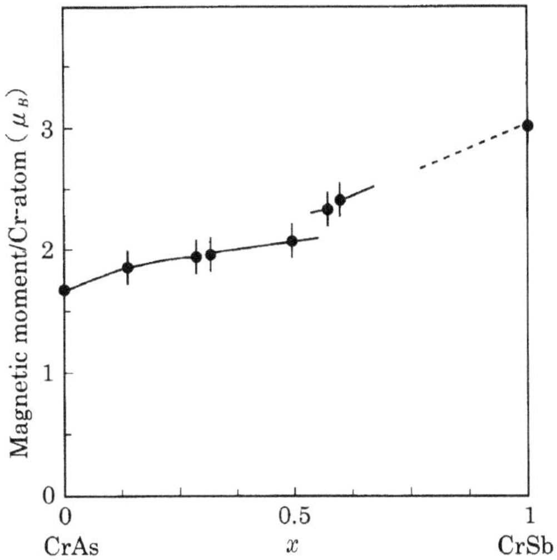

Fig. 3.44. Magnetic moment of a Cr atom in $CrAs_{1-x}Sb_x$ obtained by neutron diffraction measurements. For $x \leq 0.5$, the double helical ordering and the MnP-type structure realize (i.e., similar to CrAs). For $x > 0.5$, antiferromagnetism and the NiAs-type structure realize (similar to CrSb) [65]

3.5 CrSb

CrSb, which takes the NiAs-type structure, is an antiferromagnet with $T_N = 718\,K$ and a simple spin arrangement shown in Fig. 3.45.

Magnetic moment of a Cr atom is equal or close to Cr^{3+} moment ($3.0\mu_B$) with $g = 2$. As shown in Fig. 3.38, the a-axis of CrSb is much longer than that of CrAs, while difference of the c-axis between CrSb and CrAs is small (5.65 Å for CrSb and 5.45 Å for CrAs). Such a large a-axis of CrSb brings about narrow width of the $3d$ band and enhancement of intra-atomic exchange interaction, consequently the large moment. As we mentioned in Sect. 3.4.2, temperature dependence of susceptibility of CrSb behaves like an antiferromagnet with two sublattice [71]. This behavior is quite different from that in the temperature region $T_N \leq T \leq T_t$ for CrAs. We can obtain $\theta_P = -200\,K$ and Cr-moment $2.9\mu_B$ for CrSb from the Curie–Weiss type susceptibility together with assumption of $g = 2$. The value of $2.9\mu_B$ thus obtained is in good agreement with 3.0 and $2.7\mu_B$ by the neutron diffraction measurements shown in Fig. 3.45. As compared with CrAs, CrSb seems to have a strong localized character. This kind of tendency is also seen in Mn-pnictides such as MnBi, MnSb, MnAs, and MnP (see Fig. 3.11). The relation between a-axis and magnetic moment shown in Figs. 3.38 and 3.11 seems to suggest a way of approach to understand the magnetism of $3d$-pnictides. There is relevant discussion in Sect. 4.2 of Part II.

Resistivity of CrSb above T_N shows temperature dependence like a semiconductor [72]; however, it is unsure whether such behavior of resistivity arises from an intrinsic property of CrSb or slight crystal decomposition at high temperatures such as Sb-precipitation as seen in MnSb (Sect. 1.4 and Fig. 1.6).

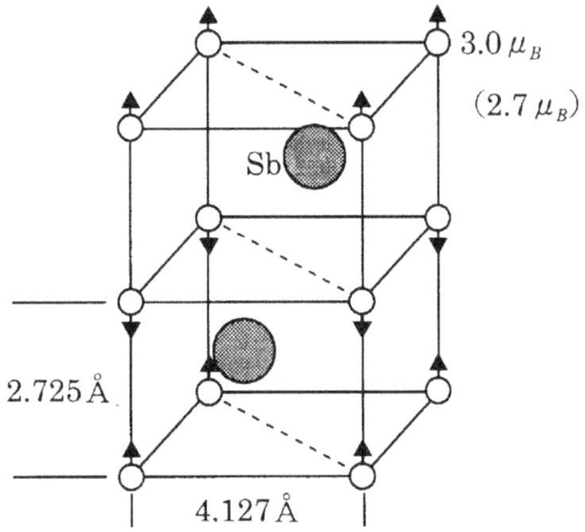

Fig. 3.45. Antiferromagnetic spin ordering of CrSb. Moment of a Cr atom has been reported as 3.0μ_B [65] or 2.7μ_B [70]

3.6 CrP

CrP takes the MnP-type structure. As shown in Fig. 3.38, the $B(a)$-axis is shorter by 0.5 Å than that of CrAs, which broadens the $3d$ band. Therefore, density of states at the Fermi level is small and no magnetic ordering appears. Susceptibility of CrP has minimum at about 200 K [73]. Magnitude and temperature dependence of the susceptibility are very similar to those of $CrAs_{0.9}P_{0.1}$ as shown in Fig. 3.36.

3.7 Properties of Cu$_2$Sb-Type Compounds

We have already shown magnetic structures of these compounds in Fig. 2.2 in Sect. 2.1. Band structures are shown in Part II of this book. In this section, we mainly discuss the very interesting magnetic properties of $Mn_{2-x}Cr_xSb$.

3.7.1 Antiferromagnetic–Ferrimagnetic Transition of $Mn_{2-x}Cr_xSb$ and Kittel's Model

It has been well known that compounds Mn_2Sb whose Mn or Sb is slightly replaced by V, Cr, Co, Cu, etc. or Ge, As etc. undergo with increase in temperature a first order transition from antiferromagnetic state in low temperature side to ferrimagnetic state in high temperature side. Among these compounds, $Mn_{2-x}Cr_xSb$ shows the most typical transition with very sharp change of magnetization. Figure 3.46 shows temperature dependence of spontaneous magnetization measured for single crystalline samples, and Fig. 3.47 shows the lattice constants as a function of temperature [74].

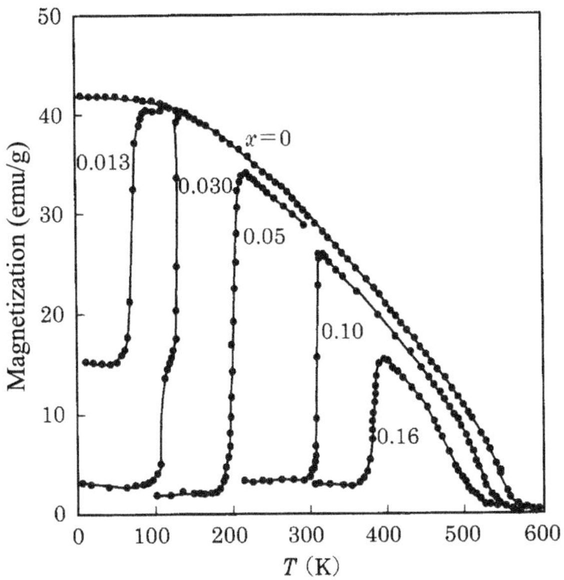

Fig. 3.46. Temperature changes of spontaneous magnetization of $Mn_{2-x}Cr_xSb$. Mn_2Sb is ferrimagnetic. Other samples with x except for $x = 0.013$ show the antiferromagnetic–ferrimagnetic transition (Reproduced from [74])

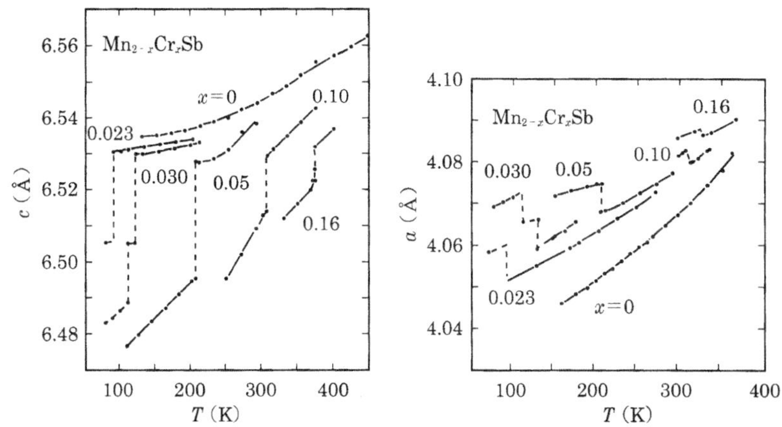

Fig. 3.47. Temperature change of lattice constants of $Mn_{2-x}Cr_xSb$ (Reproduced from [74])

In Fig. 3.46, we see the first order antiferromagnetic–ferrimagnetic transition for the samples with x in the region $0.05 \leq x \leq 0.16$, whereas we also see an intermediate magnetic ordering with a small magnetization for the samples with $x \leq 0.03$. The small magnetizations of about 3 emu g^{-1} is generally observed for all samples in the antiferromagnetic region due to small amount of impurity of ferromagnetic

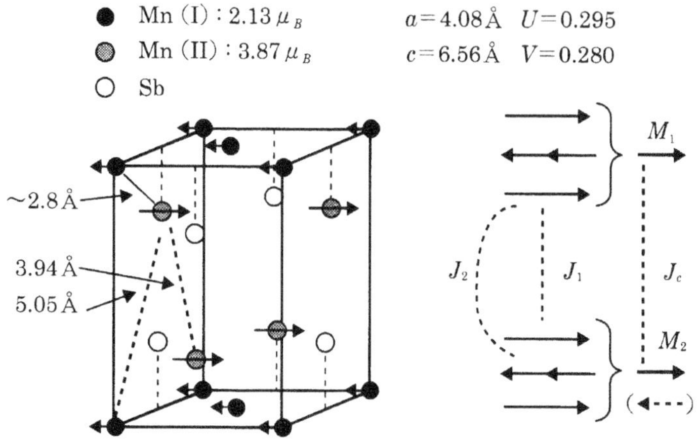

Fig. 3.48. Schematic picture of the antiferromagnetic–ferrimagnetic transition of $Mn_{2-x}Cr_xSb$. Mn atoms occupy two different sites Mn(I) and Mn(II). Moments at the Mn(I)-site and Mn(II)-site are antiparallel. M_1 and M_2 are summation of moments in three adjacent layers. When M_1 and M_2 are parallel (antiparallel), we refer it as ferrimagnetic (antiferromagnetic) state

MnSb phase. As shown in Fig. 3.47, the c-axis shows a large jump at the first order while the small jump of the a-axis may be attributed to a kind of reaction to the large jump of the c-axis.

Next we explain the change of spin alignment between the antiferromagnetic and ferrimagnetic states. The antiferromagnetic–ferrimagnetic transition can be explained by the exchange inversion model developed by Kittel [75], where the sign of exchange parameter J_c shown in Fig. 3.48 changes its sign at a critical length of the c-axis. The Kittel model is a phenomenological model, and the stability of the ferrimagnetic and the antiferromagnetic states has also been discussed from the viewpoint of itinerant electrons by Motizuki and co-workers (see Part II). We see from Figs. 3.47 and 3.48 that the c-axis length is closely related to J_c, namely it is easily seen that the contraction of the c-axis brings about the change of J_c from positive to negative values. This situation is illustrated schematically in Fig. 3.49.

We explain on the basis of the Kittel model for the antiferromagnetic-ferrimagnetic transition referring to Figs. 3.48 and 3.49. Among energies relevant to the transition, magnetic energy

$$E_{\mathrm{ex}} = -A\boldsymbol{M}_1 \cdot \boldsymbol{M}_2 \tag{3.27}$$

is directly related to the exchange parameter J_c. In this equation, $\boldsymbol{M}_1$ and $\boldsymbol{M}_2$ are sublattice magnetizations as shown in Fig. 3.48, and A is a coefficient of the molecular field given by

$$A = (2J_c Z)/(Ng\mu_{\mathrm{B}}), \tag{3.28}$$

where we note that J_c, as explained in Fig. 3.48, is an average exchange parameter of several different exchange parameters J_i connecting Mn-spins belonging to $\boldsymbol{M}_1$ and

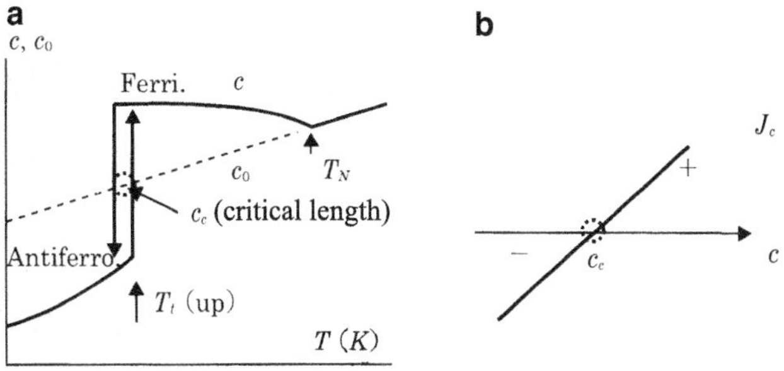

Fig. 3.49. (**a**) Temperature change of the c-axis with hysteresis is schematically shown. The experimental data in the vicinity of T_t in the process of rising temperature are shown in Fig. 3.47. The dashed line (c_0) means the normal thermal expansion curve without the spontaneous magnetostriction $(c - c_0)$. (**b**) J_c (see Fig. 3.48) plotted schematically against the c-axis length. We assume that sign of J_c changes at the critical length c_c

$\boldsymbol{M}_2$, and Z the number of Mn–Mn spin pairs. We consider here only two exchange parameters J_1 and J_2 indicated by the dashed lines in Fig. 3.48, and we have a relation $J_c Z = 4J_1 + 2J_2$. For $Z = 6$ this relation becomes $J_c = (2/3)J_1 + (1/3)J_2$. It is important that the coefficient A in (3.28) is proportional to J_c, that is, the summation of J_1 and J_2 in the form mentioned above. (Note that Kittel did not mention A in detail.) Considering that sign of J_c or A changes at $c = c_c$, we can expand A as follows,

$$A(c) = A'(c_c)(c - c_c), \tag{3.29}$$

where $A'(c_c)$ means the first derivative. Substituting (3.29) into (3.27) and replacing $A'(c_c)$ by B, we have

$$E_{\text{ex}} = -BM^2(c - c_c)\cos\theta, \tag{3.30}$$

where θ is a relative angle between $\boldsymbol{M}_1$ and $\boldsymbol{M}_2$, and $M(= M_1 = M_2)$ is magnitude of magnetization of each sublattice.

Next we consider elastic energy induced by the spontaneous magnetostriction given by the following equation:

$$\begin{aligned} E_{\text{el}} &= (1/2)[c_{11}(e_{xx}^2 + e_{yy}^2) + c_{33}e_{zz}^2] + c_{12}e_{xx}e_{yy} + c_{13}(e_{yy}e_{zz} + e_{zz}e_{xx}), \\ &= (1/2)[2c_{11}(\Delta a/a)^2 + c_{33}(\Delta c/c)^2] + c_{12}(\Delta a/a)^2 \\ &\quad + 2c_{13}(\Delta a/a)(\Delta c/c), \end{aligned} \tag{3.31}$$

where $e_{xx} = e_{yy} = \Delta a/a = (a - a_0)/a_0$ and $e_{zz} = \Delta c/c = (c - c_0)/c_0$ are shear (see Fig. 3.49a), and c_{ij}'s are stiffness constants [76]. By setting $-(\Delta a/a)/(\Delta c/c) = R(> 0)$, the elastic energy is written as

$$E_{\text{el}} = [c_{33}/2 + (c_{11} + c_{12})R^2 - 2Rc_{13}](\Delta c/c)^2 = K(c - c_0)^2, \tag{3.32}$$

with K defined by

$$K = [c_{33}/2 + (c_{11} + c_{12})R^2 - 2Rc_{13}]/c_0^2. \tag{3.33}$$

From Fig. 3.47 together with Fig. 3.49a, we see that $\Delta a/a$ is smaller than $\Delta c/c$, so that $R \simeq 0.3$. Thus, neglecting small terms in (3.33), we find K is mainly contributed by the term $Kc_{33}/2c_0^2$. Therefore, it is seen from (3.32) that the elastic energy in the present case is approximately proportional to square of magnetostriction along the c-axis. We note that different origin $c = c_c$ has been used for E_{ex} in (3.30). Magnetic entropy is unchanged during the transition at T_t, because this is a transition between different magnetic orderings.[4] We can thus omit the entropy term in free energy.

As a result, we can write the free energy as

$$F = E_{ex} + E_{el} = -BM^2(c - c_c)\cos\theta + K(c - c_0)^2. \tag{3.34}$$

From a condition $\partial F/\partial c = 0$, we have

$$c - c_0 = (B/2K)M^2\cos\theta. \tag{3.35}$$

Inserting (3.35) into (3.34), we have an expression of F which is minimum with respect to c as

$$F(\theta) = -B(c_0 - c_c)M^2\cos\theta - (B^2M^4/4K)\cos^2\theta \equiv a\cos\theta - b\cos^2\theta, \tag{3.36}$$

where $b > 0$ and a is negative (positive) for $c_0 > c_c$ ($c_0 < c_c$).

We can write as

$$F/b = D\cos\theta - \cos^2\theta \tag{3.37}$$

with

$$D = a/b = -(4K/BM^2)(c_0 - c_c). \tag{3.38}$$

Using (3.37) in Fig. 3.50, we calculate and plot F/b with various values of D as a function of θ. Equation (3.38) shows $D \leq 0$ for $c_0 \geq c_c$ and $D > 0$ for $c_0 < c_c$.

We see from this figure that $\theta = \pi$ (antiferromagnetic state) corresponds to the smallest free energy when $D > 0$ ($c_0 < c_c$) as denoted by the dotted circle. As indicated by the vertical arrow in the right hand side, D decreases as c_0 increases with rising temperature (this is normal thermal expansion plotted by the dashed line in Fig. 3.49a). At higher temperatures, D becomes negative. However, antiferromagnetic state ($\theta = \pi$) still remains stable as long as $D > -2$. It is apparent from Fig. 3.50 that a first order transition to ferrimagnetic state ($\theta = 0$) occurs at $D = -2$. Similarly, the ferrimagnetic state in the process of decreasing temperature shows a transition to the antiferromagnetic state at $D = 2$. From (3.38), we see that $c_0 = c_c + (BM^2/2K)$ at $D = -2$ and $c_0 = c_c - (BM^2/2K)$ at $D = 2$. This situation is summarized in Fig. 3.51.

[4] There is little change of entropy due to change of the molecular field.

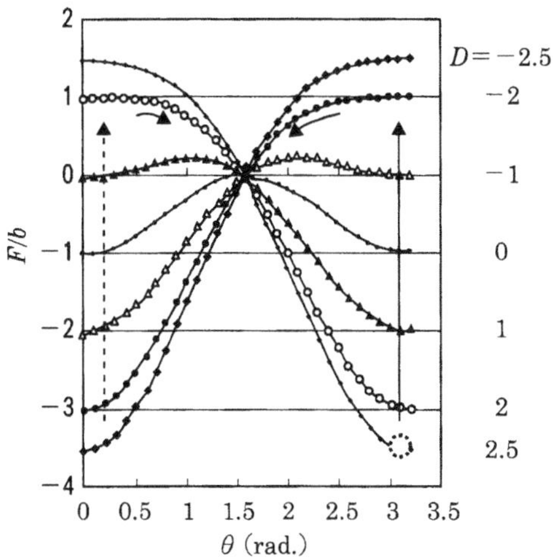

Fig. 3.50. F/b (F: free energy) plotted as a function of θ (an angle between the sublattice moments) for various values of $D(= -(4K/BM^2)(c_0 - c_c))$

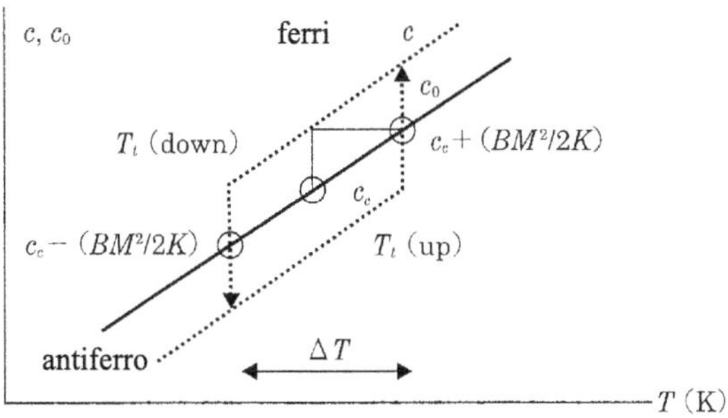

Fig. 3.51. Normal thermal expansion (c_0) and magnetostriction $(c - c_0) = \pm(BM^2/2K)$ in the vicinity of transition temperature. ΔT denotes the temperature hysteresis. Transition occurs when $c_0 = c_c \pm (BM^2/2K)$ is satisfied (see text). The *dotted lines* indicates length of the c-axis to be measured

From a relation $c_c + (BM^2/2K) = c_c[1 + \alpha(\Delta T/2)]$, ΔT is written as

$$\Delta T = (B/c_c K)M^2/\alpha, \tag{3.39}$$

where α is a thermal expansion coefficient of c_0. We see from (3.39) that width of hysteresis is proportional to M^2.

We have explained qualitatively the first order antiferromagnetic–ferrimagnetic transition of $Mn_{2-x}Cr_xSb$ in the vicinity of the first order transition temperature. Next, we compare the result mentioned above with experimental data. From (3.35), the jump of the c-axis at $T = T_t$ (up or down), $\Delta c/c$, is approximately written as

$$\Delta[(c - c_0)/c_0] \simeq \Delta[(c - c_0)/c_c] = (B/Kc_c)M^2. \tag{3.40}$$

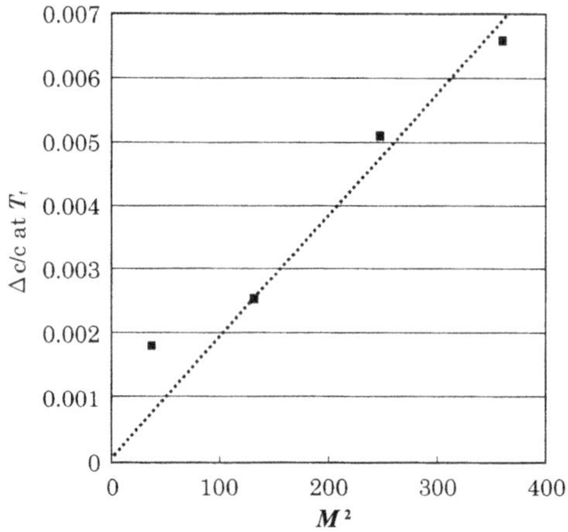

Fig. 3.52. Relation between the square of sublattice moment M^2 and the jump of the c-axis at the first order transition temperature T_t. (Drawn by using experimental data shown in Figs. 3.46 and 3.47)

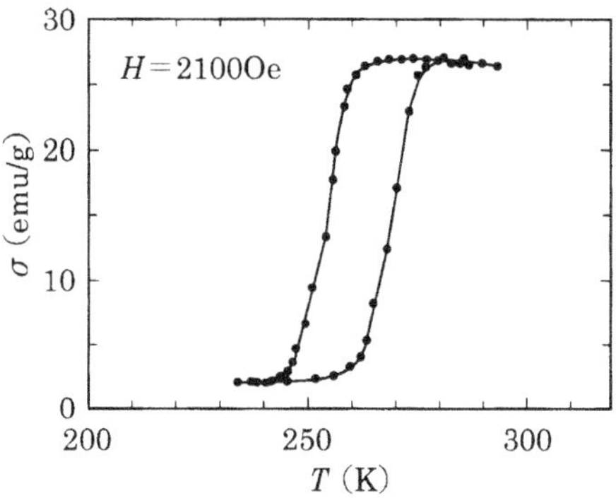

Fig. 3.53. Temperature hysteresis of magnetization for $Mn_{2-x}Cr_xSb$ ($x = 0.07$). $\Delta T \simeq 15$ K

M^2 and $\Delta(c - c_0)/c_c$, which is simply expressed by $\Delta c/c$, at $T = T_t$ can be determined from experimental results shown in Figs. 3.46 and 3.47. We note that magnetization of a sublattice M has been assumed to be a half of the jump of magnetization at T_t.

Figure 3.52 shows experimental result corresponding to (3.40). The linear relationship between $\Delta c/c$ and M^2 shown in Fig. 3.52 indicates that (3.35) is satisfied. From the slope of the dotted line in Fig. 3.52 and (3.35), we can evaluate (B/Kc_c) as 1.9×10^{-5}. Applying this value to (3.39), we obtain

$$\Delta T = 1.9 \times 10^{-5} M^2/\alpha. \tag{3.41}$$

Now we show temperature variation of magnetization in Fig. 3.53 to show an example of the hysteresis ΔT. Note that it is somewhat difficult to determine the correct value of ΔT from experiments, because the hysteresis depends on the rate of temperature change and the sample quality, etc. In addition, a large coexistence region of low temperature antiferromagnetic phase and high temperature ferrimagnetic phase has been found by X-ray diffraction measurements.

From Fig. 3.53, we see that $\Delta T \simeq 15\,\text{K}$. Inserting $\Delta T = 15\,\text{K}$ and $M^2 \simeq 150\,(\text{emu/g})^2$ (see Fig. 3.46) in (3.41), we can estimate as $\alpha \simeq 2.5 \times 10^{-4}\ \text{K}^{-1}$ though direct measurement of α is not available. This value of α is larger than α of a similar compound $Mn_2Sb_{0.875}As_{0.125}$ [77], however, usual one in the intermetallic compounds.

We note that there is no other model to explain the first order antiferromagnetic–ferrimagnetic transition of $Mn_{2-x}Cr_xSb$. The model described above is based on the assumption that the exchange parameter J_c connecting different sublattices is sensitive to length of the c-axis and changes its sign at the critical length c_c. As shown in Fig. 3.49, J_c is the summation of J_1 and J_2. If J_1 and J_2 have different sign and comparable magnitude, sign of J_c may change due to even small change of the c-axis. It is noted that J_1 and J_2 mean exchange interactions between rather distant Mn atoms; the nearest distance is 5.05 Å for J_1 and 3.94 Å for J_2 as shown in Fig. 3.48. It is expected to investigate J_1 and J_2 on the basis of band structures.

3.7.2 Magnetic Transition of $Fe_{a-x}Mn_xAs$ ($a \simeq 2$)

As for compounds written as $Fe_{a-x}Mn_xAs$, both Fe_2As and Mn_2As show antiferromagnetic ordering. For $a = 2.1$ and $1.25 \leq x \leq 1.5$, however, ferrimagnetic structure occurs in low temperature region. This ferrimagnetism undergoes a first order transition to antiferromagnetic ordering with rising temperature [78].

3.7.3 Layered Ferromagnets MnAlGe and MnGaGe

These compounds have an easy axis of magnetization along the c-axis. In these compounds Mn atoms occupy only the Mn(I)-site shown in Fig. 3.48, and Al (Ga) and Ge occupy the Mn(II)-site and the Sb-site, respectively. Mn atoms thus align within a two-dimensional plane (001), and the distance between the neighboring planes is as long as the c-axis (5.933 Å for MnAlGe). These compounds therefore have a magnetically two-dimensional character. Probably the RKKY (Ruderman–Kittel–Kasuya–Yoshida) interaction and super-exchange interaction via Al, Ga, and Ge are responsible for the magnetic interaction between Mn atoms in the adjacent layers. Curie temperature T_C and magnetic moment per $3d$ atom in $Mn_{1-x}M_xAlGe$ are plotted against x, respectively, in Figs. 3.54 and 3.55 ($M = 3d$ metals) [79,80]. By making use of the band of MnAlGe [81], we can explain the reduction of moment for the case of M = Fe, which has one more electron within the rigid band picture. The case of Cr substitution is also explained in the similar way. The prominent increase of T_C by 80 K due to 15% Cr substitution, which is shown in Fig. 3.54, contrasts to that the Cr-substitution decreases the averaged moment of $3d$ atom. Such increase of T_C may be caused by change of the inter-layer interaction as well as intra-layer interaction. MnGaGe shows properties similar to MnAlGe [82]. Band structures and magnetic properties of these compounds are also discussed in Part II of this book.

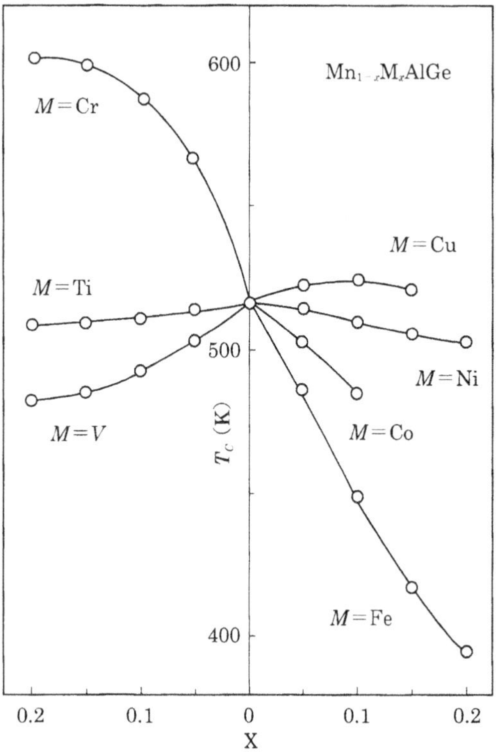

Fig. 3.54. T_C of $Mn_{1-x}M_xAlGe$ (M = 3*d* atom) plotted as functions of composition (Reproduced from [79])

3.7.4 Application of the First Order Transition of $Mn_{2-x}Cr_xSb$

As shown in Fig. 3.53, the hysteresis of the antiferromagnetic–ferrimagnetic transition is rather large. A composite material made of compounds $Mn_{2-x}Cr_xSb$ with two slightly different Cr compositions is expected to show a magnetization–temperature curve as shown schematically by solid bold lines in Fig. 3.56.

To explain this figure, let us start from an antiferromagnetic state at the initial temperature T_0, which is not shown in the figure, however, between T_2' and T_1. When the temperature is raised, the initial magnetization M_0 is zero at temperatures $T < T_1$, and then the nonmagnetic state changes discontinuously to the magnetic state with residual magnetization M_1 when temperature is raised up to T_1. M_1 remains unchanged in the temperature range of $T_1 < T < T_2$. With further heating, residual magnetization M_1 similarly increases to M_2.

It is thus possible to store three values M_0, M_1, and M_2 at one position (bit) by different three heating temperatures T's with $T_2' < T < T_1$, $T_1 < T < T_2$, and $T_2 < T$. Thus, a composite material made of the three compounds with different Cr-composition can, in principle, store four values. It may be possible to store more data using a composite material made of compounds with continuously different

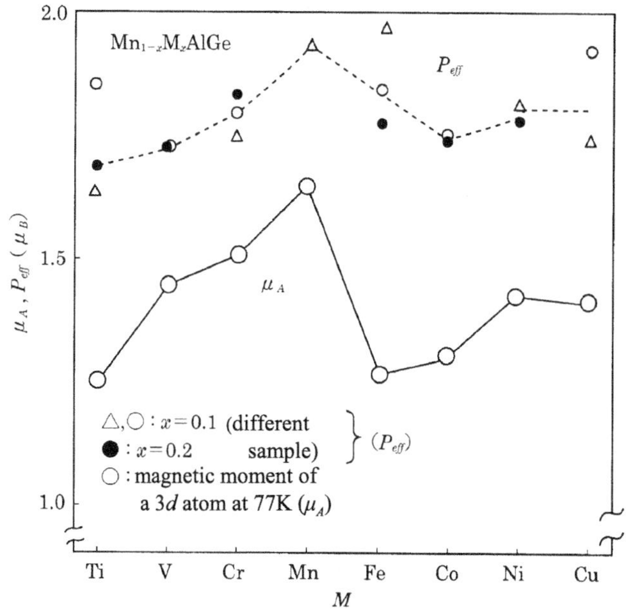

Fig. 3.55. Magnetic moment per 3*d* atom in $Mn_{1-x}M_xAlGe$ (M = 3*d* atom). P_{eff} is effective moment estimated from the Curie–Weiss constant, and μ_A is magnetic moment evaluated from spontaneous moment (Reproduced from [79])

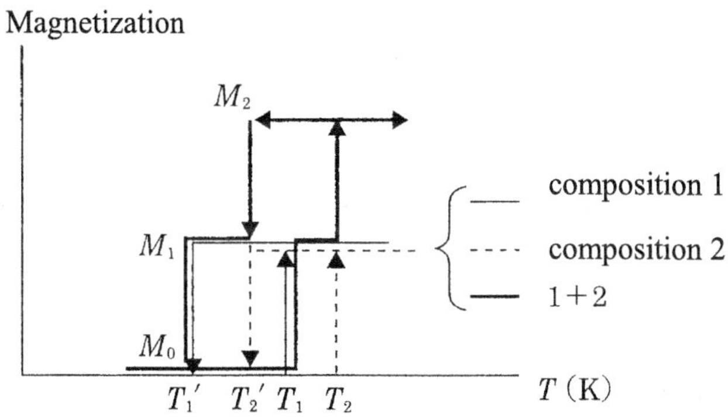

Fig. 3.56. Magnetization–temperature curves of composite material made of two compounds having different Cr compositions. T_1 and T_2 (or T_1' and T_2') are the first order transition temperatures of the compound in heating and cooling processes, respectively. $(T_1 - T_1')$ etc. is temperature width of hysteresis

Cr compositions. This mechanism is superior to an usual magnetic memory, which stores two values at a position (bit) corresponding to plus/minus magnetization. We need to pay attention that the hysteresis curves in Fig. 3.56 are idealized. Actual magnetization curves show jump of magnetization within a finite temperature range as shown in Figs. 3.53 and 3.46. It is necessary to reduce this temperature range by, for example, homogenization of Cr composition [83]. This compound system $Mn_{2-x}Cr_xSb$ may be applicable easily for temperature controller because large variety of transition temperatures can be obtained by changing the Cr composition.

References

[1] T. Komatsubara, T. Suzuki, E. Hirahara, J. Phys. Soc. Jpn. **28**, 317 (1970)
[2] Y. Shapira, C.C. Becerra, N.F. Oliviera Jr., T. Chang, Phys. Rev. **24B**, 2780 (1981)
[3] T. Komatsubara, H. Shinohara, E. Hirahara, J. Appl. Phys. **40**, 1037 (1969)
[4] A. Isizuka, T. Komatsubara, E. Hirahara, J. Phys. Soc. Jpn. **30**, 292 (1971)
[5] Landolt-Börnstein III/27a, *Magnetic Properties of Pnictides and Chalcogenides*, ed. by K. Adachi, S. Ogawa (Springer, Berlin, 1989), p. 70
[6] G.P. Felcher, J. Appl. Phys. **37**, 1056 (1966)
[7] E.E. Huber, D.H. Ridgley, Phys. Rev. A **135**, 1033 (1964)
[8] H. Ido, J. Magn. Magn. Mater. **70**, 205 (1970)
[9] N. Iwata, H. Fujii, T. Okamoto, J. Phys. Soc. Jpn. **46**, 778 (1979)
[10] H. Fjellvag, A. Kjekshus, A. Zieba, S. Foner, J. Phys. Chem. Solids **45**, 709 (1984)
[11] H. Fjellvag, A. Kjekshus, Acta. Chem. Scand. A **38**, 703 (1984)
[12] H. Fjellvag, A. Kjekshus, A.F. Andresen, Acta Chem. Scand. A **39**, 143 (1985)
[13] C. Guillaud, H. Crevedux, C.R. Acad. Sci. Paris **224**, 266 (1947)
[14] H. Ido, J. Appl. Phys. **57**(part IIA), 3247 (1985)
[15] T. Suzuki, H. Ido, J. Phys. Soc. Jpn. **51**, 3149 (1982)
[16] H. Ido, T. Harada, K. Sugiyama, T. Sakakibara, M. Date, *High Field Magnetism*, ed. by M. Date (North-Holland, Amsterdam, London, 1983), p.175
[17] C.P. Bean, D.S. Rodbell, Phys. Rev. **126**, 104 (1962)
[18] R.W. DeBlois, D.S. Rodbell, Phys. Rev. **130**, 1347 (1963)
[19] N.P. Grazhdankina, E.A. Zavadskii, I.G. Fakidov, Sov. Phys. Solid State **11**, 1879 (1970)
[20] A. Zieba, Y. Shapira, S. Foner, Phys. Lett. A **91**(5), 243 (1982)
[21] H. Ido, T. Suzuki, I. Iguchi, J. Magn. Magn. Mater. **31–34**, 159 (1983)
[22] O. Nashima, T. Suzuki, H. Ido, K. Kamishima, T. Goto, J. Appl. Phys. **79**(8), 4647 (1996)
[23] N. Menyuk, J.A. Kafalas, K. Dwight, J.B. Goodenough, Phys. Rev. **177**, 942 (1969)
[24] N. Kazama, H. Watanabe, J. Phys. Soc. Jpn. **30**, 1319 (1971)
[25] K. Selte, A. Kjekshus, P. Peterzens, A.F. Andresen, Acta Chem. Scand. A **32**, 653 (1978)
[26] K. Bärner, C. Santandrea, V. Neitzel, E. Gmelin, Phys. Status Solids **45**, 541 (1984)
[27] H. Fjellvag, A. Kjekshus, Acta Chem. Scand. A **38**, 1 (1978)
[28] S. Haneda, N. Kazama, Y. Yamaguchi, H. Watanabe, J. Phys. Soc. Jpn. **42**, 1212 (1977)
[29] L.H. Schwartz, E.L. Hall, G.P. Felcher, J. Appl. Phys. **42**, 1621 (1971)
[30] G. Bödecker, K. Bärner, K. Funke, Phys. Status Solidi (B) **98**, 571 (1980)
[31] H. Fjellvag, A.F. Andresen, K. Barner, J. Magn. Magn. Mater. **46**, 29 (1984)
[32] J.B. Goodenough, D.H. Ridgley, W.A. Newman, Proc. Int. Conf. Magnetism, Nottingham (1964) p. 542
[33] H. Krokoszinski, C. Santandrea, E. Gmelin, K. Barner, Phys. Status Solidi **113**, 185 (1982)
[34] T. Suzuki, H. Ido, J. Phys. Soc. Jpn. **51**, 3149 (1982)
[35] H. Ido, J. Phys. Soc. Jpn. **25**, 1543 (1968)
[36] H. Ido, T. Suzuki, J. Magn. Magn. Mater. **104–107**, 1939 (1992)
[37] K. Sugiyama, I. Shiozaki, H. Ido, M. Date, Physica B **155**, 303 (1989)

[38] G. Kido, H. Ido, J. Magn. Magn. Mater. **70**, 207 (1987)
[39] H. Ido, S. Yasuda, M. Kido, G. Kido, T. Miyakawa, J. de Physique C8 **167** (1988)
[40] H. Ido, S. Yasuda, G. Kido, J. Appl. Phys. **69**, 4621 (1991)
[41] K. Bärner, Phys. Status Solidi (A) **5**, 1699 (1971)
[42] H. Nagasaki, I. Wakabayashi, S. Minomura, J. Phys. Chem. Solids **30**, 329 (1969)
[43] H. Nagasaki, I. Wakabayashi, S. Minomura, J. Phys. Chem. Solids **30**, 2405 (1969)
[44] H. Yamada, Phys. Rev. B **47**, 11211 (1993)
[45] E.A. Zavadskii, B. Todris, Sov. Phys. Solid State **18**, 173 (1976)
[46] T. Goto, M.I. Batashevich, K. Kondo, K. Terao, H. Yamada, H. Ido, J. Alloys Comp. **325**, 18 (2001)
[47] G.A. Samara, A.A. Giardini, *Physics of Solids at High Pressure*, ed. by C.T. Tomizuka, E.M. Emrick (Academic, New York, 1961), p. 308
[48] B.T.M. Willis, H.P. Rooksby, Proc. Phys. Soc. **67**, 290 (1954)
[49] Y. Takahashi, J. Phys. Condens. Matter **2**, 8405 (1990)
[50] H. Ido, J. Magn. Magn. Mater. **70**, 205 (1987)
[51] T. Kamimura, H. Ido, S. Sato, T. Suzuki, J. Magn. Magn. Mater. **54–57**, 939 (1986)
[52] S. Shimotomai, H. Ido, J. Appl. Phys. **99**, 08109 (2006)
[53] H. Wada and Y. Tanabe, Appl. Phys. Lett. **79**, 3302 (2001)
[54] H. Wada, K. Taniguchi, Y. Tanabe, Mater. Trans. JIM **43**, 73 (2002)
[55] F. Gronvold, S. Snildal, E. Westrum Jr., Acta Chem. Scand. **24**, 285 (1970)
[56] S.M. Benford, G.V. Brown, J. Appl. Phys. **52**, 2110 (1981)
[57] T. Chen, W. Stutius, IEEE Trans. Magn. **10**, 581 (1974)
[58] M. Kishimoto, K. Wakai, J. Appl. Phys. **48**, 4640 (1977)
[59] K. Selte, A. Kjekshus, W.A. Jamison, A.F. Andresen, J.E. Engebresen, Acta Chem. Scand. **25**, 1703 (1971)
[60] K. Selte, H. Hjersing, A. Kjekshus, A.F. Andresen, P. Fischer, Acta Chem. Scand. A **29**, 695 (1975)
[61] T. Suzuki, H. Ido, J. Magn. Magn. Mater. **54–57**, 935 (1986)
[62] N. Kazama, H. Watanabe, J. Phys. Soc. Jpn. **30**, 1319 (1971)
[63] N. Kazama, H. Watanabe, J. Phys. Soc. Jpn. **31**, 943 (1971)
[64] K. Bärner, C. Santandrea, V. Neitzel, E. Gmelin, Phys. Status Solidi (B) **123**, 541 (1984)
[65] A. Kallel, H. Boller, E.F. Bertaut, J. Phys. Chem. Solids **35**, 1139 (1974)
[66] G.P. Felcher, F.A. Smith, D. Bellavance, A. Wold, Phys. Rev. B **39**, 3046 (1971)
[67] K. Selte, A. Kjekshus, W.A. Jamison, A.F. Andresen, J.E. Engebresen, Acta Chem. Scand. **35**, 1042 (1971)
[68] T. Suzuki, H. Ido, J. Appl. Phys. **73**, 5686 (1993)
[69] C. Kittel (ed.) *Introduction to Solid State Physics*, 2nd edn (Wiley, New York), p. 154
[70] A.I. Snow, Rev. Mod. Phys. **25**, 127 (1953)
[71] K. Adachi, K. Sato, K. Ohmori, C. Ito, T. Ido, Toyoda Kenkyu Hokoku **24**, 64 (1971)
[72] T. Suzuoka, J. Phys. Soc. Jpn. **12**, 1344 (1957)
[73] K. Selte, L. Birkeland, A. Kjeksjus, Acta Chem. Scand. A **32**, 731 (1978)
[74] F.J. Darnell, W.H. Cloud, H.S. Jarrett, Phys. Rev. **130**, 647 (1963)
[75] C. Kittle, Phys. Rev. **120**, 335 (1960)
[76] C. Kittel (ed.) *Introduction to Solid State Physics*, 2nd edn. (Wiley, New York), p. 134
[77] K. Shirakawa, H. Ido, J. Phys. Soc. Jpn. **40**, 666 (1976)
[78] T. Kanomata, T. Goto, H. Ido, J. Phys. Soc. Jpn. **43**, 1178 (1977)
[79] H. Ido, T. Kamimura, K. Shirakawa, J. Appl. Phys. **55**, 2365 (1984)
[80] T. Kamimura, H. Ido, K. Shirakawa, J. Appl. Phys. **57**(Part IIA), 3255 (1985)
[81] See Part II of this book.
[82] J.B. Goodenough, G.B. Street, K. Lee, J.C. Suits, J. Chem. Solids **36**, 451 (1975)
[83] N. Takahashi, S. Shimotomai, H. Ido, J. Appl. Phys. **97**, 10M513 (2005)

Appendix: Magnetic Transition and Free Energy

Free energy G of a magnetic material can be written in the Landau type expansion form as

$$G = G(0) + a_2\sigma^2 + a_4\sigma^4 + a_6\sigma^6 + \ldots, \tag{A1}$$

where σ is a magnetic moment of a ferromagnet (or a sublattice moment for an antiferromagnet) normalized by the value at $T = 0\,\mathrm{K}$. When an external magnetic field is applied, Zeeman energy $-\sigma H$ is added. This expansion form is irrelevant to the nature of electrons, localized or itinerant; however, physical meaning of the coefficients depends on the character of electrons and theoretical model. In this Appendix, we show an expansion form of magnetic entropy, which appears in the Bean–Rodbell theory of magnetic phase transition of MnAs, up to the sixth order σ^6. Further, we show expression of $G(\sigma)$ and its temperature dependence.

Let us consider a paramagnetic material consisting of N particles with spin j in a magnetic field H. In this case free energy is written as $G = -gj\mu_\mathrm{B}\sigma HN - TS(\sigma)$. Replacing $S(\sigma)$ by $Nk_\mathrm{B}S(\sigma)$, we have $G/Nk_\mathrm{B}T = -(gj\mu_\mathrm{B}\sigma/k_\mathrm{B}T)\sigma - S(\sigma) = -\alpha\sigma - S(\sigma)$. Applying the minimum condition $\partial G/\partial\sigma = 0$ to the final equality, we obtain

$$\alpha = -\partial S(\sigma)/\partial\sigma. \tag{A2}$$

On the other hand, α is related to σ through the Brillouin function with spin j, $B_j(\alpha)$, as

$$\sigma = B_j(\alpha) = A\alpha + B\alpha^3 + D\alpha^5 + \ldots, \tag{A3}$$

where $A, B, D, \ldots$ are given by

$$A = (1/3)[(2j+1)^2 - 1]/(2j)^2, \quad B = -(1/45)[(2j+1)^4 - 1]/(2j)^4,$$
$$D = (1/945)[(2j+1)^6 - 1]/(2j)^6 \ldots. \tag{A4}$$

By expanding $S(\sigma)$ in (A2) with respect to σ, we have

$$\begin{aligned} S(\sigma) &= S(0) + a\sigma^2 + b\sigma^4 + c\sigma^6 + \cdots \\ &= \ln(2j+1) + a\sigma^2 + b\sigma^4 + c\sigma^6 + \cdots \end{aligned} \tag{A5}$$

As the number of microscopic states, in the case of $\sigma = 0$, is given by $Z = (2j+1)^N$ and $S = k_B \ln Z$, it is reasonably understood that $S(0)$ becomes $\ln(2j+1)$. By inserting (A5) into (A2), we have $\alpha = -(2a\sigma + 4b\sigma^3 + 6c\sigma^5 + \cdots)$. By inserting this relation into (A3) and setting the coefficient of each σ^n term zero, we have the expansion coefficients $a, b, c, \ldots$ as

$$\begin{aligned} &a = -1/A < 0, \quad b = B/4A^4 < 0, \\ &c = -(1/2)(B^2/A^7) + D/6A^6 < 0, \end{aligned} \tag{A6}$$

where A, B, and D are given by (A4). We insert (A5) into the expression of magnetic entropy (3.2) of Sect. 3.2.1, and then we put a condition of a minimum energy with respect to volume $\partial G/\partial V = 0$ into (3.2) again. Thus, we obtain the free energy G also minimized by crystal volume. By setting $T/T_0 = t$ (the relative temperature), $P = 0$, and $H = 0$, we have

$$G(\sigma)/Nk_BT_0 = -t\ln(2j+1) - a(t-1)\sigma^2 - b(t-\eta)\sigma^4 - ct\sigma^6 + \cdots, \tag{A7}$$

with

$$\eta = -(a^2/2b)Nk_BKT_0\beta^2 > 0. \tag{A8}$$

Figure 3.57 displays $[G(\sigma) - G(0)]/Nk_BT_0$ calculated by (A7) with parameters $\eta = 2$ and $j = 3/2$, which were chosen so as to be applicable to MnAs.

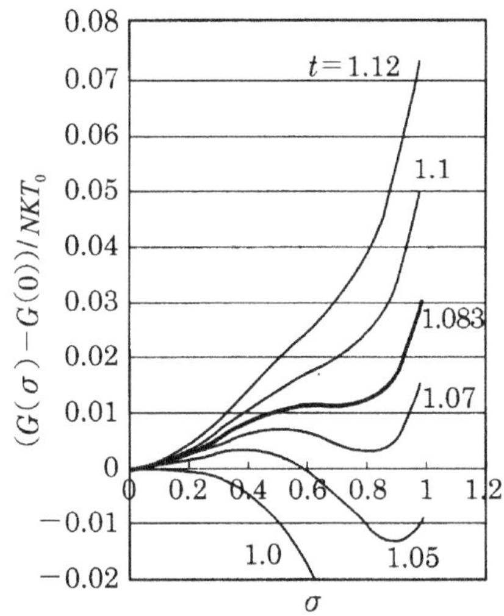

Fig. 3.57. Free energy with parameters $j = 3/2$ and $\eta = 2$ vs. normalized moment σ drawn for various relative temperatures

On heating, the minimum point of free energy disappears at temperature $t = 1.083$ as seen in Fig. 3.57, which means that magnetization σ disappears discontinuously due to the absence of energy barrier. Therefore, this is of the first order and also illustrated in the σ-T curve with $\eta = 2$ in Fig. 3.5. While in the process of decreasing temperature from about $t = 1.083$, the peak of free energy disappears at $t = 1$. This correspond to sudden appearance of magnetization. We note that the curves in Fig. 3.57 are somewhat inaccurate in the region $\sigma \simeq 1$ because of truncation of the expansion formula of entropy. (Numerical calculation is necessary to evaluate entropy without the expansion form.) The coefficient of the σ^4-term of free energy has an important meaning. When $\eta \geq 1$, this coefficient is negative in the temperature range $1 \leq t \leq \eta$. As coefficients of σ^2 and σ^6 terms are positive for $t \geq 1$, the free energy does not have minimum at finite σ unless the coefficient of the σ^4 term is negative. Therefore, the first order transition takes place only for $\eta \geq 1$.

Part II

Itinerant Electron Theory

4

Electronic Band Structure and Magnetism of NiAs-Type Compounds

4.1 Band Calculation

Extensive studies on band structures of NiAs-type transition-metal compounds started in the middle of 1980s by the group of Motizuki, who is one of the present authors, of Osaka University and the group of Haas of the University of Groningen.

For the band calculations, the group of Osaka University employed the self-consistent augmented plane wave (APW) method, while the group of the University of Groningen used the self-consistent augmented spherical wave (ASW) method. In spite of the difference in the method used for calculations, the general features of calculated bands are similar to each other. In the APW calculations carried out by the group of Osaka University, muffin-tin approximation was used for atomic potential. For exchange and correlation interaction, the local (spin) density approximation (LDA or LSDA) was adopted within the formula of Gunnarsson and Lundqvist (GL). The equation of von Barth and Hedin (HB) for the LDA was also used in some calculations. In this book, we show mainly the results calculated by the group of Osaka University.

In recent years, we have applied more sophisticated methods. For example, we carried out band calculations by using the self-consistent full-potential linearized augmented plane wave (FLAPW) method to obtain more accurate band structures. Kulatov and his co-workers carried out band calculations by using the self-consistent linear-muffin-tin orbital (LMTO) method. Nakada and Yamada recently calculated bands by using the LMTO method with atomic sphere approximation (ASA); this method which employs localized basis functions is convenient for the calculations for a system including many atoms in the unit cell.

Scalar relativistic effects are usually included, however, spin–orbit interaction (SOI) is neglected in many band calculations. The SOI removes the degeneracy in the dispersion curves and also induces mixture between up-bands and down-bands, resulting in reduction of magnetic moment of ferromagnetic state. Furthermore, the SOI in the band calculations is needed to investigate optical processes. Recently, the SOI has been included in the calculations by Kulatov et al.

Generalized gradient approximation (GGA) has been used to improve the LDA. The GGA has enabled us to see difference of total energies between different structures. For example, total energy calculation for CoAs having the NiAs- and the MnP-type structures has been carried out by Nakada, and his results by the full-potential LMTO method with the GGA show good agreement with experiments.

4.2 Band Structures and Optical Properties

4.2.1 Pnictides: MnAs and MnSb

Nonmagnetic Band of MnAs and MnSb with NiAs-Type Structure

Figure 4.1a, b shows the crystal structure and the first Brillouin zone of the NiAs-type structure, respectively. Figure 4.2a shows the dispersion curves of nonmagnetic MnAs calculated by the self-consistent APW method with the muffin-tin approximation and the LDA [1]. The two bands in the low-energy region consist of As-4*s* orbitals. Above the gap, there are 16 mixing bands consisting of As-4*p* and Mn-3*d* orbitals. Figure 4.2b shows the density of states (DOS) of the *p*–*d* mixing bands. It is seen from this figure that there are two remarkable peaks: the low-energy region is contributed by bonding orbitals of As-4*p* and Mn-3*d*, and the peaks in high-energy region by antibonding orbitals mainly consisting of 3*d* states of Mn. As seen in Fig. 4.2b, the width of the antibonding *p*–*d* mixing bands is about 5 eV (or 0.4 Ry), which suggests that *d* states of MnAs are not localized but are of itinerant character. Density of states of MnSb has features similar to those of MnAs, however the band width is slightly smaller than that of MnAs.

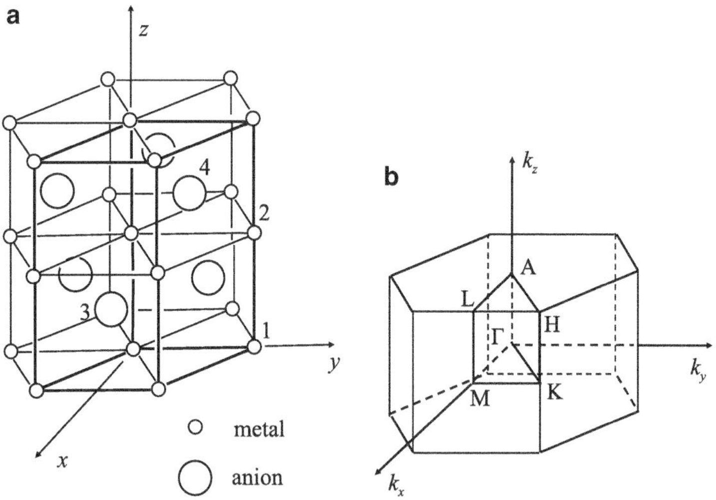

Fig. 4.1. (**a**) The NiAs-type crystal structure and (**b**) the first Brillouin zone

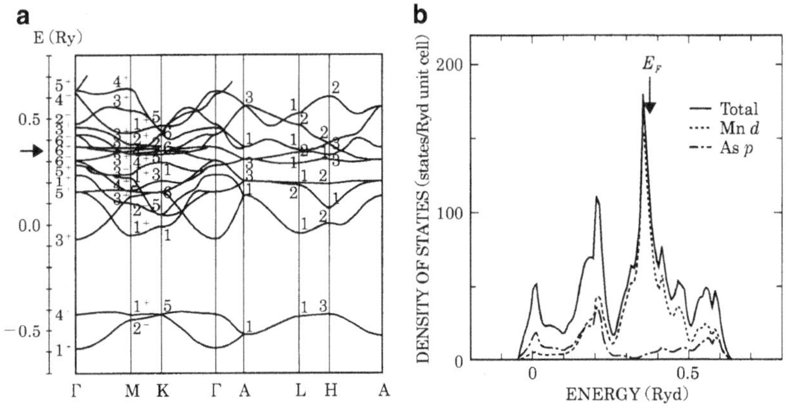

Fig. 4.2. (**a**) Dispersion curves and (**b**) density of states of the nonmagnetic MnAs calculated by the APW method

As shown in Fig. 4.2b, the Fermi level is near the large peak of the density of states, which is contributed mainly by the d orbitals of Mn. The density of states at E_F, $\rho(E_F)$, of MnAs is larger than that of NiAs-type compounds with $3d$ elements other than Mn, which is consistent with the fact that both MnAs and MnSb are ferromagnetic below the Curie temperatures $T_C = 318$ K and 537 K, respectively.

These densities of states calculated by the APW method agree well with the bands of MnAs and MnSb calculated by Sandratskii by using the Green's function method [2]. Recently, Nakada and Yamada have calculated nonmagnetic bands of MnAs having the NiAs-type structure by the FLAPW method with the GL formula and the LMTO–ASA method with the BH formula. Their results shown in Fig. 4.3a, b are rich in fine structure, however the general features of the bands are similar to the bands obtained by using the APW method shown in Fig. 4.2b.

Ferromagnetic Bands of MnAs and MnSb with the NiAs-Type

Among the NiAs-type compounds, only MnAs, MnSb, and MnBi are ferromagnetic as shown in Part I. We have calculated ferromagnetic bands using the self-consistent APW method. The muffin-tin approximation for atomic potential and the LSDA within the GL formula for the exchange and correlation interaction were used.

The density of states of the p–d mixing bands of ferromagnetic MnAs is shown in Fig. 4.4. Due to spin-polarization, the mixing bands of As-$4p$ and Mn-$3d$ orbitals split in the up-spin and the down-spin subbands, whereas no polarization occurs in the $4s$ bands in the low-energy region. In Fig. 4.4, the broken and dot-dashed curves show components of Mn-$3d$ and As-$4p$ orbitals in the muffin-tin spheres, respectively. Comparing Fig. 4.4 with the nonmagnetic bands shown in Fig. 4.2, we see that the p–d mixing bands are not shifted rigidly, but deformed significantly due to the spin-polarization.

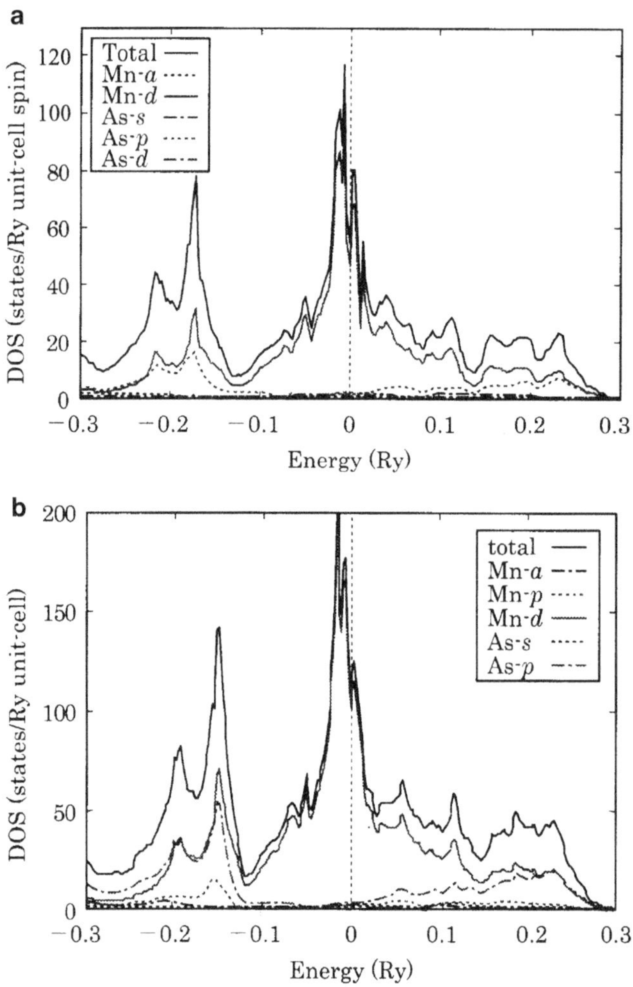

Fig. 4.3. Density of states of nonmagnetic MnAs calculated using the (**a**) FLAPW method and (**b**) LMTO–ASA method (K. Nakada, Dr. thesis)

We evaluated total magnetic moment from the calculated band as $3.1\mu_B$/formula. This value is somewhat smaller than the observed magnetic moment $3.4\mu_B$/formula (see Part I). Magnetic moment in the muffin-tin spheres of a Mn-site and an As-site is $3.12\mu_B$/formula and $-0.15\mu_B$/formula, respectively, hence most of the magnetic moment is attributed to Mn atoms. In addition to the magnitude of the total magnetic moment, the calculated values of magnetic moments at Mn- and As-sites correspond well to the results of neutron diffraction measurements, therefore we see that magnetic moments in MnAs are well understood from the band theory.

Ferromagnetic band of MnSb with the NiAs-type structure is similar to the band of MnAs except that the width of *p*–*d* mixing band is slightly smaller than

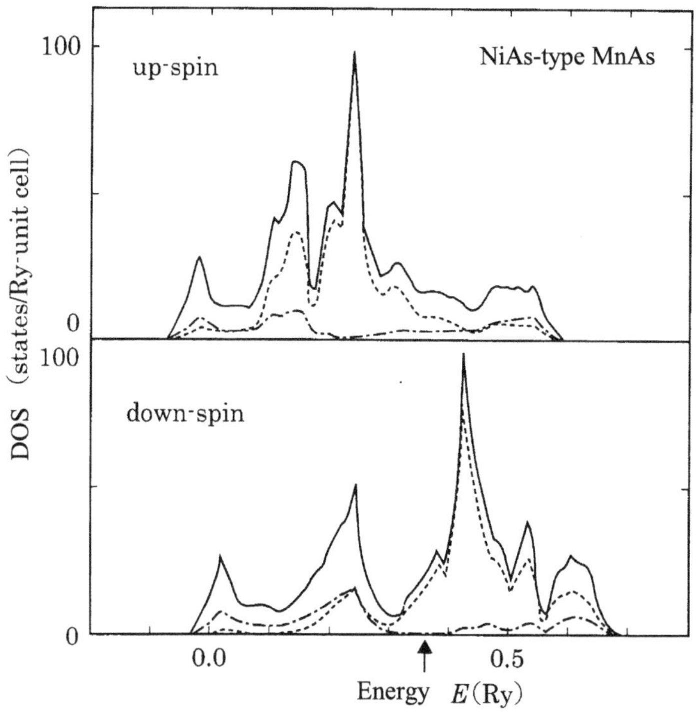

Fig. 4.4. Density of states of ferromagnetic MnAs calculated by the APW method. The *broken* and *dash-dotted* lines are Mn-3d and As-4p components, respectively

that of MnAs. According to the band calculation, the total magnetic moment is 3.2μ_B/formula; magnetic moment inside the muffin-tin sphere at Mn- and Sb-sites are 3.41μ_B/formula and $-0.12\mu_B$/formula, respectively. These values of magnetic moment agree fairly well with those measured by Yamaguchi et al. [3].

Results of our band calculation for MnSb agree well with the bands calculated by Coehoorn et al. by using the self-consistent ASW method [4], namely both dispersion curves and density of states in these two works agree well with each other, and the magnetic moments by Coehoorn et al. are 3.3μ_B at a Mn-site and $-0.06\,\mu_B$ at an Sb-site. Density of states at the Fermi level $\rho(E_F)$ is obtained by us to be 37.0 (states/Ry unit cell), which agrees well with 35.8 (states/Ry unit cell) calculated by Coehoorn et al. These values are also in good agreement with 32.6 (states/Ry unit cell) evaluated from measurements of specific heat.

X-ray photoelectron spectroscopy (XPS) measurements for MnSb, CoSb, and NiSb were carried out by Liang and Chen [6]. Coehoorn et al. compared the XPS spectrum with their ferromagnetic bands (the 5s band and the p–d mixing band) of MnSb. As shown later in Fig. 4.17 in Sect. 4.2.5, there is good correspondence for the peaks of the XPS spectrum and density of states (see also Sect. 4.2.5).

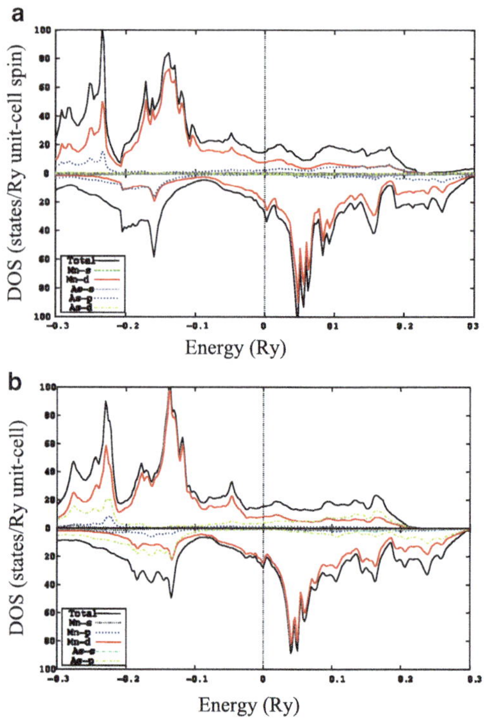

Fig. 4.5. Density of state of ferromagnetic MnAs. (**a**) FLAPW method, (**b**) LMTO-ASA method [5]

Nakada and Yamada calculated ferromagnetic band of MnAs with the NiAs-type structure by using both the FLAPW method and the LMTO–ASA method [5]. Calculated densities of states are shown in Fig. 4.5a, b. Compared with the band obtained by the APW method, these densities of states are rich in fine structures, however, their general features are similar to those calculated by the APW method.

Nakada and Yamada also evaluated contributions of five Mn-*d* orbitals and three As-*p* orbitals to the APW wavefunctions. Their results are shown in Fig. 4.6, where five *d*-components for density of states and their values (or numbers of electron belonging to the respective components) are depicted as a function of electron energy. Contributions of five *d*-components to charge and magnetic moment are also shown in Table 4.1, where the values are normalized so that the summation of contributions from the five *d* orbitals is 100%. These calculations revealed that Mn-*d* orbitals are not extended in the *c*-plane, but have rather spherical distribution. They also plotted charge distribution in the (0001) plane and (1$\bar{1}$00) plane which include Mn–Mn bonds. The results also confirmed that charge density of Mn-*d* orbitals has spherical character. Furthermore, they clarified by plotting charge density in the (11$\bar{2}$0) plane including Mn and As atoms that the Mn–As bond is the strongest bond. The charge distributions are shown in Fig. 4.7.

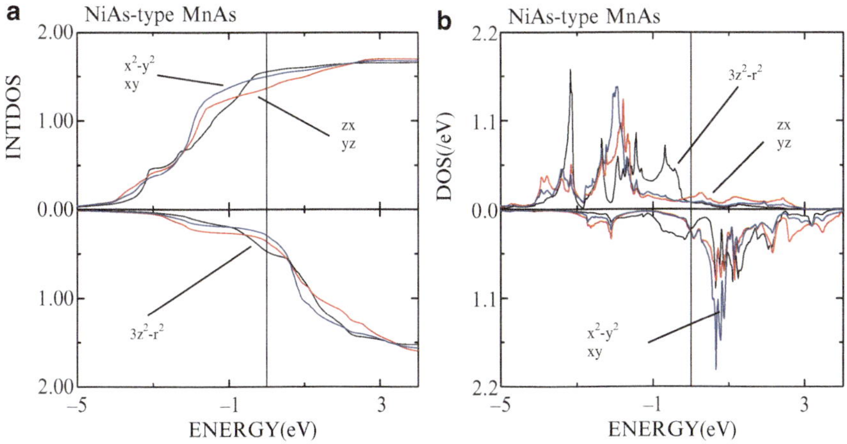

Fig. 4.6. Density of states of ferromagnetic MnAs calculated by the FLAPW method. (**a**) Integrated DOS and (**b**) components of each *d* orbitals (K. Nakada, private communication)

Table 4.1. Contribution rates of *d* orbitals for charge and magnetic moment, which are integrated up to the Fermi level, of MnAs

	$3z^2-r^2$	zx	zy	x^2-y^2	xy
Charge (%)	22	19	19	20	20
Moment (%)	20	18	18	22	22

They also investigated the effect of crystal volume change on magnitude of magnetic moment. They carried out band calculations in two cases: one is the case of fixed-axial ratio c/a (i.e., the bond angle Mn–As–Mn is kept constant) and another is the case of unfixed-axial ratio (i.e., the bond angle is variable). They found that magnitude of magnetic moment is significantly affected by the change of bond angle. This result indicates that magnetic moment is sensitive to lattice distortion in the direction of the *a*-axis. On the other hand, effect of lattice distortion along the *c*-axis on magnetic moment is small, though Mn–Mn distance in this direction is smaller than the distance along the *a*-axis. These results of band calculation are consistent with the experimental data seen, for example, in Figs. 3.3 and 3.11.

Nonmagnetic Band of MnAs with MnP-Type Structure

MnAs shows a structural transformation to the MnP-type structure at $T_t = 318$ K (see also Fig. 3.2). We calculated band structure of MnAs with the MnP-type structure by using the self-consistent APW method. The band structure deforms due to the lattice distortion accompanied with the structural transformation, which results in decrease of $\rho(E_F)$. In Sect. 4.3, we will show in detail that anomalous behaviors of

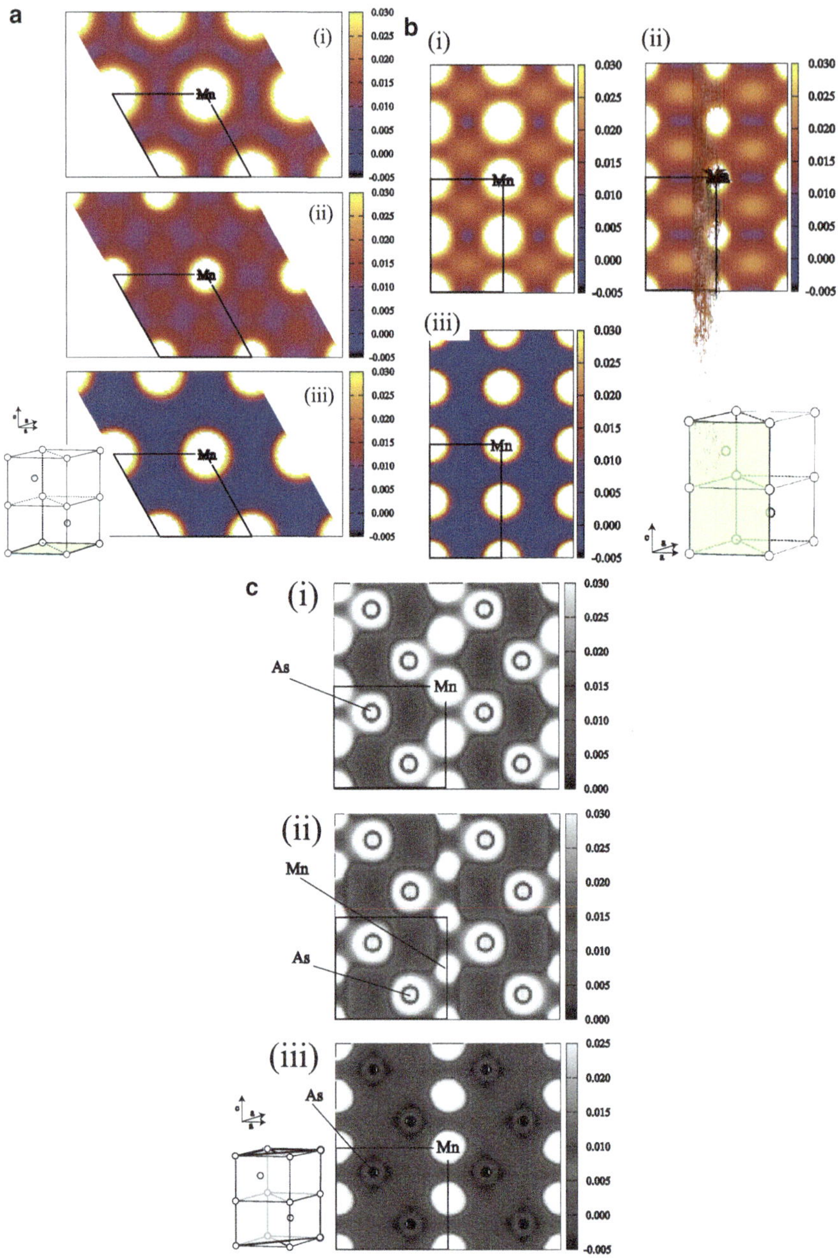

Fig. 4.7. Charge distribution of ferromagnetic MnAs (**a**) (0001) plane, (**b**) $(1\bar{1}00)$ plane, (**c**) $(11\bar{2}0)$ plane. In each figure, (i), (ii), and (iii) show majority spin, minority spin, and moment, respectively [5]

paramagnetic susceptibility and thermal expansion observed for MnAs (see Fig. 3.2) are well explained by considering spin fluctuation effect and the deformation of the band structure.

4.2.2 FeAs, CoAs, and NiAs

It is known that CoAs shows structural transformation at $T_t = 1,250$ K from the MnP-type structure (low-temperature phase) to the NiAs-type structure (high-temperature phase). FeAs has the MnP-type structure, and NiAs has the NiAs-type structure.[1] FeAs has the double-helical magnetic ordering lower than $T_N = 77$ K, whereas CoAs and NiAs are nonmagnetic at all temperatures (see Table 2.7).

Nonmagnetic Bands of CoAs and NiAs with the NiAs-Type Structure

We calculated band structures of CoAs and NiAs with the NiAs-type structure by using the self-consistent APW method within the muffin-tin potential approximation and the LDA. Density of states of the *p–d* mixing band of CoAs is shown in Fig. 4.8. Density of states at the Fermi level is $\rho(E_F) = 46$ (states/Ry unit cell) for CoAs and 31 (states/Ry unit cell) for NiAs. These values are much smaller than 156 (states/Ry unit cell) of MnAs, which is consistent with the fact that CoAs and NiAs are nonmagnetic but MnAs is ferromagnetic.

We also depicted Fermi surfaces of CoAs and NiAs, and then found that CoAs has two hole surfaces around the A-axis and these surfaces are expected to have good nesting for a displacement of wavevector ΓM. Since this wavevector ΓM is relevant

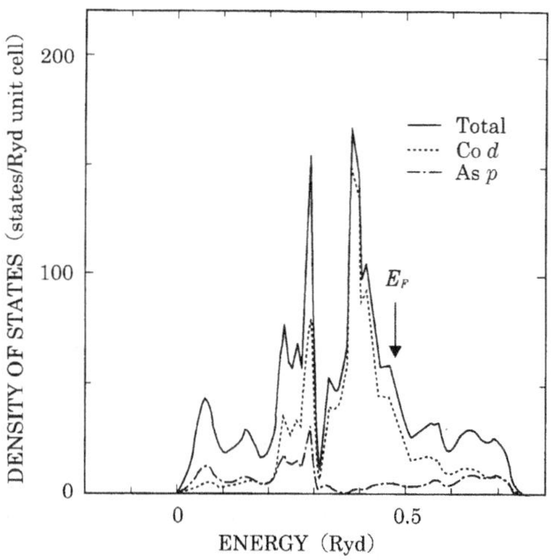

Fig. 4.8. Density of states of nonmagnetic CoAs with the NiAs-type structure calculated by the APW method

[1] We have related argument in Sect. 4.4.

to deformation into the MnP-type structure, the NiAs-type structure of CoAs has an instability to the MnP-type structure. This fact corresponds well to the high T_t of CoAs. On the other hand, the structural transformation to the MnP-type structure will not occur for NiAs because no nesting of Fermi surfaces is expected [7]. We have related discussion in Sects. 4.4 and 4.7.

Nonmagnetic Bands of FeAs and CoAs with the MnP-Type Structure

Figure 4.9 shows DOS curves for the *p–d* mixing band of nonmagnetic CoAs and FeAs with the MnP-type structure. These compounds have characteristic DOS curves, namely, the densities of states show abrupt increase just below the Fermi level [7]. We will show in Sect. 4.3 that anomalous temperature dependence of paramagnetic susceptibility of CoAs and FeAs can be well explained by the spin fluctuation theory including such feature of the densities of states.

It is known that total energies evaluated from band structures make it possible to know which crystal structure is most stable at 0 K. Recently, Nakada and coworkers calculated nonmagnetic bands of CoAs with the NiAs-type and the MnP-type structures by using the full-potential LMTO method with the GL formula for the exchange and correlation interaction. From the calculated total energies they found that the MnP-type structure is more stable than the NiAs-type structure for CoAs. This result is consistent with observation that CoAs remains in the MnP-type structure up to 1,250 K. They obtained also similar result by calculation of full-potential LMTO method with the GGA term.

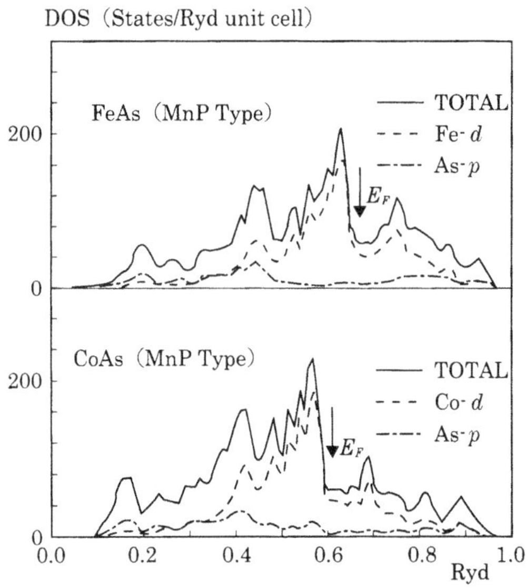

Fig. 4.9. DOS curves calculated by the APW method for nonmagnetic CoAs and FeAs having MnP-type structure

4.2.3 CrSb, CrAs, and CrP

These compounds take the NiAs-type or the MnP-type structure [8]. As shown in Table 4.2, the lattice constants of these compounds generally decrease as the anion changes in the order Sb, As, and P. In particular, change of the *a*-axis is remarkable. The CrX compounds show various magnetic properties with variation of anion atom. CrSb is an antiferromagnet with Cr moment of $3.0\mu_B$ [9, 10]. CrAs takes a double-helical magnetic ordering, and the magnetic moment of Cr is $1.7\mu_B$, which is only about a half of Cr-moment of CrSb [11]. CrP is a paramagnetic compound whose magnetic susceptibility is almost independent on temperature [12]. The double-helical magnetic ordering of CrAs has been known to disappear by a few percent substitution of P atoms for As [13]. The experimental results mentioned earlier suggest that there is a close correlation between the crystal structure (or lattice parameters) and magnetic properties in the CrX compounds.

To investigate the correlation between crystal structure and magnetic properties of the CrX compounds, we carried out first-principle band calculations by using the full potential linear muffin-tin orbital (FP–LMTO) method [14]. The crystallographic parameters in Table 4.2 were used in the calculations. As for the local density approximation (LDA) for exchange and correlation interaction, we have used the formula of Vosko, Wilk, and Nussair [15]. We carried out iterative calculations using 125 $\boldsymbol{k}$-points in the 1/24 of the first Brillouin zone so that muffin-tin potential converges within 10^{-3} mRy.

Figure 4.10 shows densities of states of nonmagnetic CrP, CrAs, and CrSb. From these densities of states we see following three features:

1. Total band width increases in the order CrSb, CrAs, and CrP.
2. Hybridization between Cr $3d$-orbitals and anion orbitals takes place in both high- and low-energy regions, but not in the intermediate region.
3. Width of the large DOS peak near the Fermi level in CrSb, where there is little Cr–X mixing broadens and splits roughly in two groups as X goes from Sb to P.

Table 4.2. Crystallographic parameters of CrX (X = P, As, or Sb) at the room temperature. Lattice constants *a*, *b*, and *c* are common to both NiAs- and MnP-type structures (see Sect. 1.3). *u*, *v*, *w*, and *x* denote displacement parameters [8]

	CrP	CrAs	CrSb
Crystal structure	MnP-type	MnP-type	NiAs-type
a (Å)	3.114	3.463	4.127
b (Å)	6.018	6.212	7.148
c (Å)	5.360	5.649	5.451
u	0.05	0.05	0
v	0.065	0.05	0
w	0.018	0.006	0
x	0.0073	0.0065	0

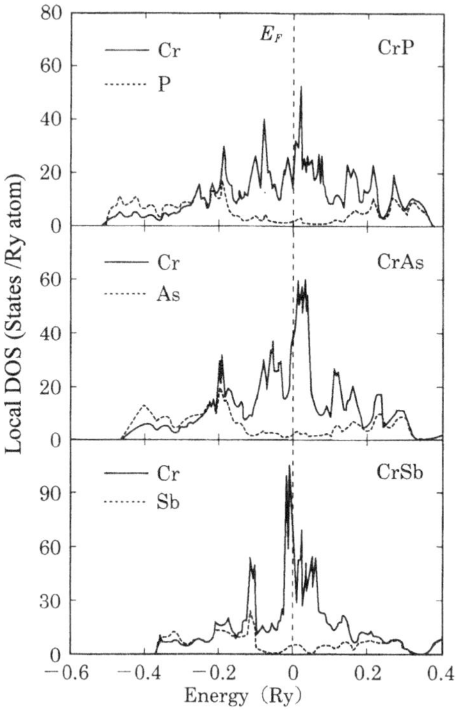

Fig. 4.10. Local densities of states for CrP, CrAs, and CrSb (/Ry atom)

As shown in Table 4.2, the smallest distance between Cr atoms in the c-direction, which is equal to $c/2$ ($\sim$2.7Å) in the case of the NiAs-type structure, is almost common to all the CrX compounds. On the other hand, length of the a-axis which corresponds to the distance between Cr atoms within the c-plane, depends considerably on X of the CrX compounds. This difference in the length of the a-axis influences the bonding character in the c-plane, resulting in the characteristics of the density of states especially in the vicinity of E_F as mentioned earlier.

We evaluated contribution of five d-orbitals to the total density of state. The results for CrSb and CrAs are shown in Fig. 4.11. As shown in Table 4.2, CrSb and CrAs take different crystal structure. The length of the a-axis of CrAs is much smaller than that of CrSb. We discuss relation between such difference of the a-axis length and magnetic properties. As shown in Fig. 4.11, the density of states of CrSb has a sharp peak near the Fermi level. This part consists of mainly $d(xy)$ and $d(x^2-y^2)$ components which extend in the c-plane and do not mix with p states of Sb as shown in Fig. 4.10. On the other hand, states near the Fermi level of CrAs consist of not only $d(xy)$ and $d(x^2-y^2)$ but also $d(xz)$ and $d(yz)$ orbitals. Because of the short a-axis of CrAs compared with that of CrSb, width of $d(xy)$- and $d(x^2-y^2)$-bands become larger for CrAs, and then the Fermi level is pushed down to the low-energy

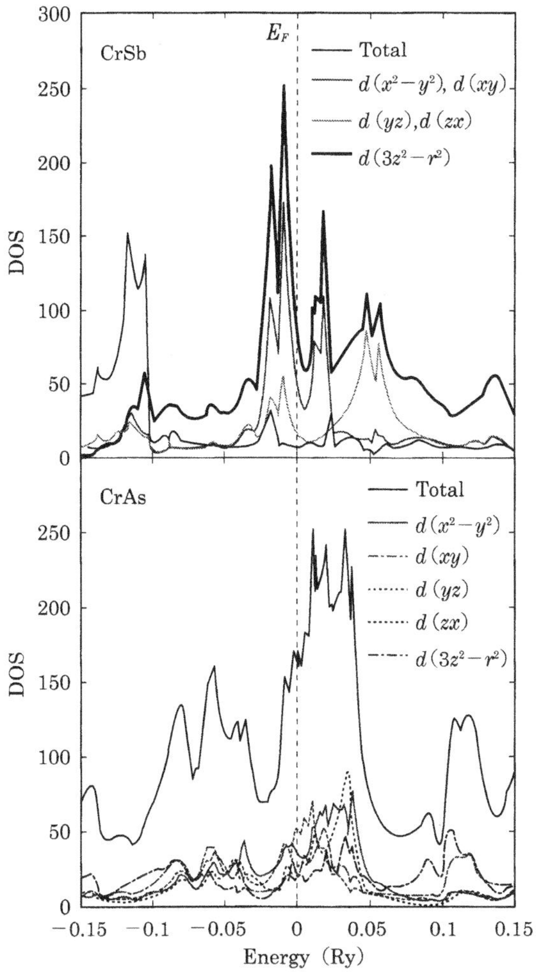

Fig. 4.11. Contribution of 3d orbitals on density of states of CrSb and CrAs

side of the peak of the density of states. The atomic displacement of As which is denoted by v in Table 4.2 due to the MnP-type distortion gives rise to lowering the energy of $d(xz)$ and $d(yz)$ components, however, these states remain in the energy region higher than the Fermi level. The low-energy (high-energy) peak is bonding (antibonding) band consisting of $d(xz)$, $d(yz)$, $d(3z^2 - r^2)$ states and p state of anion as seen in Fig. 4.10.

It is important that the change of the a-axis, which induces change of $d(xy)$ and $d(x^2 - y^2)$ bands, plays an important role to explain the variety of magnetic properties seen in CrSb, CrAs, and CrP.

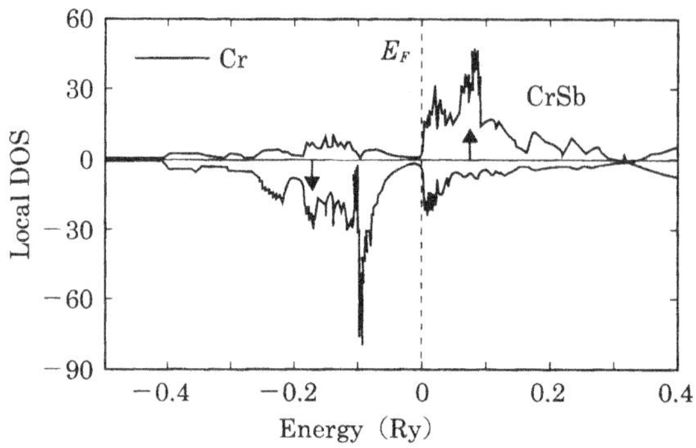

Fig. 4.12. Density of states of Cr for one of the two magnetic sublattices of antiferromagnetic CrSb (in unit of states/Ry atom)

From the density of states of CrP shown in Fig. 4.10, $\rho(E_F)$ is evaluated as 27.3 (states/Ry f.u.), which is 1/2 and 1/4 of the value of $\rho(E_F)$ estimated from measurements of specific heat [16] and magnetic susceptibility measurement [17], respectively.

Figure 4.12 shows density of states for one of the two magnetic sublattices of antiferromagnetic CrSb, which gives Cr magnetic moment of $2.7\mu_B$. This value agrees well with experimental values 2.7 μ_B [10] and 3.0 μ_B [9]. Since there is little *p–d* mixing in the region near the Fermi level as shown in Fig. 4.10, there is strong polarization of Cr-*d* states; most of the up-spin states lies above the Fermi level. This fact indicates that Cr-*d* states in CrSb has a considerable localized character probably because of the large length of *a*-axis. It is also notable that the down-spin band has a gap-like shape near the Fermi level.

Calculated total energies for paramagnetic, ferromagnetic, and antiferromagnetic states of CrSb are plotted in Fig. 4.13 as a function of crystal volume. As seen in this figure, the band calculation indicates that the antiferromagnetic state is most stable. This is consistent with observation that CrSb is an antiferromagnet.

To understand magnetic properties of CrAs (double helix, $\mu_{Cr} = 1.7\mu_B$, $T_N = 250$ K, see Sect. 3.4 for detail), it is expected to investigate wavevector-dependent susceptibility $\chi(\boldsymbol{q})$. This is an important problem left for the future.

4.2.4 Chalcogenides: CrTe, CrSe, and CrS

Compounds between Cr and a chalcogen (S, Se, or Te) also take the NiAs-type structure. CrTe is a ferromagnet with $T_C = 340$ K, whereas CrSe and CrS take antiferromagnetic ordering below $T_N = 285$ K and 460 K, respectively. Effects of high pressure on these compounds have attracted much interest. From electron-spin resonance (ESR) measurements, Shanditsev pointed out that ferromagnetism of CrTe disappears at the critical pressure 2.8 GP [18]. Kanomata and his co-workers investigated pressure dependence of Curie temperature T_C and magnetic moment [19]. They pointed out that the ferromagnetic state is not the ground state under high

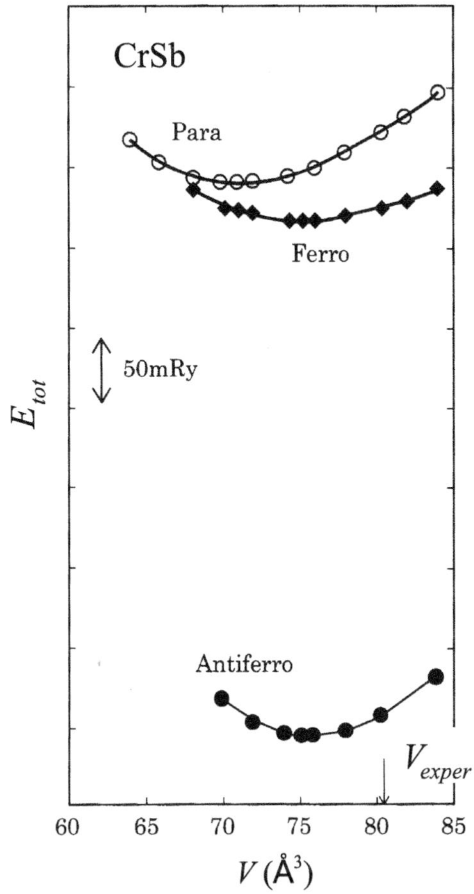

Fig. 4.13. Total energy of CrSb as a function of unit cell volume. The total energies for paramagnetic state, ferromagnetic state, and antiferromagnetic state are plotted. V_{exper} denotes experimental value of unit cell volume

pressure. According to observation by Ishiduka and Eto, the ferromagnetism of CrTe disappears at 7 GP and a structural transition to the MnP-type structure occurs at 13 GP [20].

To explain the results of these high-pressure experiments, we have theoretically studied the possible magnetic transitions by investigating pressure dependence of band structures. We will discuss the pressure effect in detail in Sect. 4.5, and in this section we describe the results of band structures calculated by Nakada and co-workers; they investigated dependence of bond length (Cr–Cr) and bond angle (Cr–Te–Cr) on charge distribution and magnetic moment in ferromagnetic CrTe.

On the basis of the FLAPW band, calculations for ferromagnetic CrTe, Nakada decomposed the APW wavefunction into the components from three p orbitals of Te and five *d* orbitals of Cr. The results are shown in Fig. 4.14a, b and Table 4.3. The density of states and integrated density of states are shown in Fig. 4.14b and 4.14a, respectively.

Contribution from each *d* orbital to charge and magnetic moment in CrTe is shown in Table 4.3, where the values are normalized so that the summation of con-

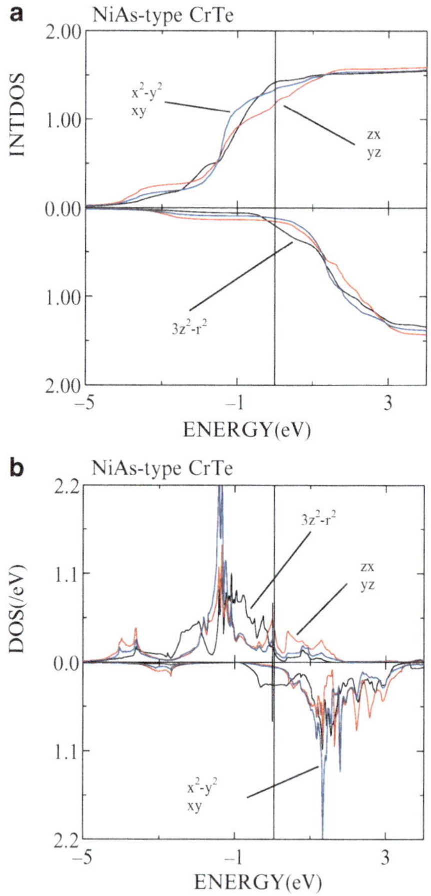

Fig. 4.14. (**b**) Density of state of ferromagnetic CrTe calculated by the FLAPW method. (**a**) Integrated DOS component from each *d* orbital (K. Nakada, private communication)

Table 4.3. Contribution rates of respective *d* orbitals to charge and magnetic moment in CrTe

	$3z^2 - r^2$	zx	zy	$x^2 - y^2$	xy
Charge (%)	22	19	19	20	20
Moment(%)	21	18	18	21.5	21.5

tributions from five d orbitals is 100%. We see that all d orbitals have almost equal contribution to charge and magnetic moment, which means that charge of the d states of a Cr atom distributes spherically as also seen in MnAs. This result that charge distribution of Cr is spherical has been confirmed by depicting charge distribution in the (0001) and $(1\bar{1}00)$ planes which include Cr–Cr bond. They also found strong Cr–Te bond from charge distribution in the $(11\bar{2}0)$ plane which include Cr and Te. They have clarified that the strongest bond between Cr and Te is the bond in the $(11\bar{2}0)$ plane.

They also investigated the effect of volume change on magnetic moment. They calculated magnetic moment as a function of volume for the two cases: (i) bond angle of Cr–Te–Cr is fixed (i.e., the axial ratio c/a is fixed) and (ii) bond angle of Cr–Te–Cr is unfixed (i.e., c/a is variable). The calculated results indicate that magnetic moment is sensitive to lattice change along the a-axis where Cr–Cr distance is large, and insensible to lattice change along the c-axis where Cr–Cr distance is small. This is a situation similar to that of MnAs.

4.2.5 Optical Properties

We can discuss optical properties of compounds on the basis of band calculation including the spin–orbit interaction.

Kulatov and co-workers calculated ferromagnetic bands of MnAs and MnSb by using the LMTO method including the spin-orbit interaction [21]. In the calculations, they used the muffin-tin approximation and the expression of von Barth and Hedin (BH) for the exchange and correlation interactions, and they also took the spin–orbit interaction into account.

Figure 4.15a, b shows the calculated densities of states of MnAs and MnSb, respectively. Except the split of the band due to the spin–orbit interaction, overall

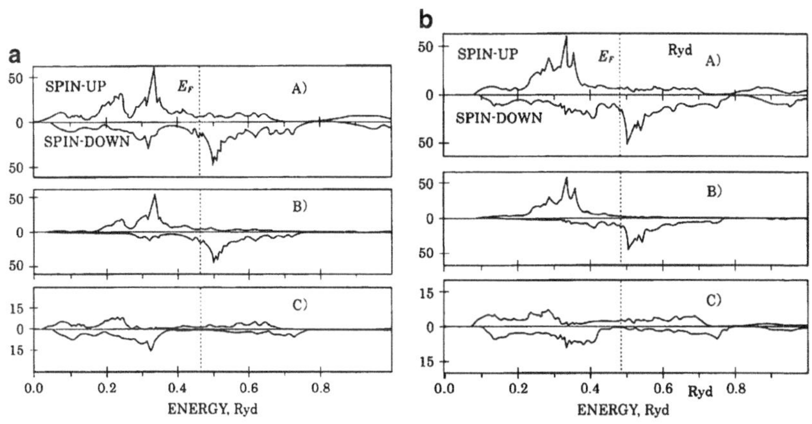

Fig. 4.15. Densities of states of (**a**) ferromagnetic MnAs and (**b**) MnSb. (A) shows total DOS. (B) and (C) show partial DOS of Mn-d and As(Sb)-p, respectively (Reproduced from [21])

features of the bands are similar to the bands calculated by Motizuki and her co-workers with the self-consistent APW method. Magnetic moment evaluated here by the LMTO method agrees well with experimental value.

On the basis of these bands, Kulatov investigated optical properties of MnAs and MnSb. He obtained a dielectric function by making use of the Kubo formula for the linear response theory. Optical absorption coefficient is proportional to the imaginary part $\varepsilon_2(\omega)$ of the dielectric function, and $\varepsilon_2(\omega)$ consists of two terms: the Drude term $\varepsilon_2^D(\omega)$ arising from intra-band transition, and the $\varepsilon_2^b(\omega)$ term arising from interband transition. In the dipole approximation, $\varepsilon_2^b(\omega)$ is given by

$$\varepsilon_2^b(\omega) = \frac{8\pi^2 e^2}{3m^2\omega^2\Omega} \sum_{\boldsymbol{k}} \sum_{\lambda' \neq \lambda} |\langle \boldsymbol{k}\lambda' | \boldsymbol{j} | \boldsymbol{k}\lambda \rangle|^2 f_{\boldsymbol{k}\lambda'}(1 - f_{\boldsymbol{k}\lambda})\, \delta(E_{\boldsymbol{k}\lambda'} - E_{\boldsymbol{k}\lambda} - \hbar\omega), \quad (4.1)$$

where $f_{\boldsymbol{k}\lambda}$ is the Fermi–Dirac distribution function, $\boldsymbol{j} = -\mathrm{i}\frac{e\hbar}{m}\boldsymbol{\nabla}$, and λ is a band suffix. The matrix elements for the dipole transition $\langle \boldsymbol{k}\lambda' | \boldsymbol{j} | \boldsymbol{k}\lambda \rangle$ were calculated from the LMTO wavefunctions. Interband optical photoconductivity $\sigma(\omega)$ is related to $\varepsilon_2^b(\omega)$ as

$$\varepsilon_2^b(\omega) = \frac{4\pi}{\omega}\sigma(\omega). \quad (4.2)$$

Figure 4.16a, b shows the photoconductivity $\sigma(\omega)$ calculated for ferromagnetic MnAs and MnSb, where magnetic moments are assumed to align along the c-axis (z-axis). In these figures, $\sigma_{xx}(\omega)$ and $\sigma_{zz}(\omega)$ are plotted by the solid and broken curves, respectively. Due to anisotropy arising from the hexagonal symmetry of the crystal structure, shape and peak positions are different for σ_{xx} and σ_{zz}.

The features of $\sigma(\omega)$ of MnAs and MnSb in Fig. 4.16a, b are similar even in the structures. The main peaks of $\sigma(\omega)$ are attributed to transition of electrons as follows:

- The peak at around $\hbar\omega \simeq 0.3$ eV is attributed to direct interband transition which has become allowed owing to mixing between the up-spin and down-spin bands due to the spin–orbit interaction.
- The peaks at around $\hbar\omega \simeq 1$ eV correspond to interband transition between down-spin bands of Mn.
- The broad and strong peak in the energy region of $\hbar\omega = 2.5 \sim 4.5$ eV is mainly due to transition between p-band of Sb (As) and d-band of Mn.

In Fig. 4.17, we show XPS spectrum of MnSb measured by Liang and Chen (by the dashed curve) [4], and the calculated density of states of ferromagnetic MnSb below Fermi level (the solid curve). Since the measurements were carried out with a poly-crystal sample, the experimental spectrum corresponds to $\sigma_{zz} + 2\sigma_{xx}$. There is rough correspondence between the experiment and the calculation, however, experimental data on single-crystal are necessary for detailed comparison.

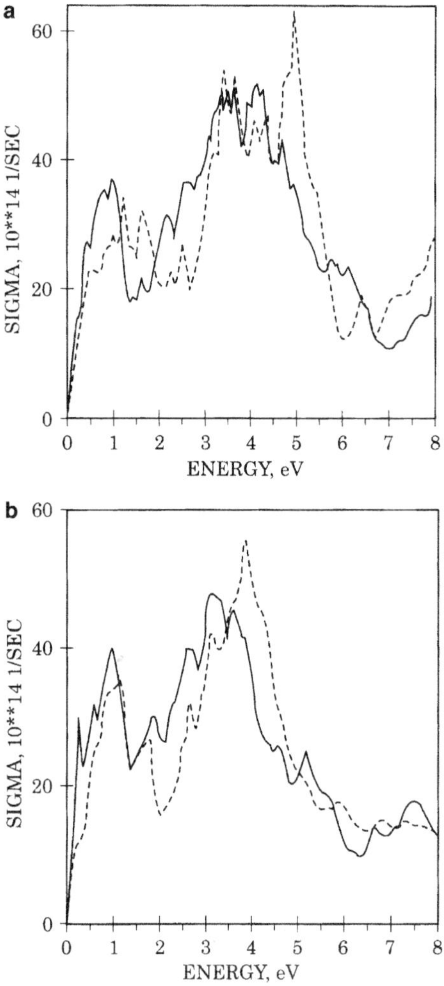

Fig. 4.16. Optical conductivity for (**a**) ferromagnetic MnAs and (**b**) ferromagnetic MnSb calculated by Kulatov. The solid and dashed curves show σ_{xx} and σ_{zz}, respectively

4.3 Spin Fluctuations and Anomalous Magnetic and Elastic Properties

4.3.1 Paramagnetic Susceptibility and Anomalous Thermal Effect of MnAs and $MnAs_{1-x}P_x$

With increasing temperature, MnAs shows a first-order transition from ferromagnetic state to paramagnetic state at $T_C = 318$ K. [2] At this temperature, the structural transformation from the NiAs-type to the MnP-type structure also occurs with 2% of

[2] See Sect. 3.2 for experimental detail of MnAs, MnP and their mixed compounds.

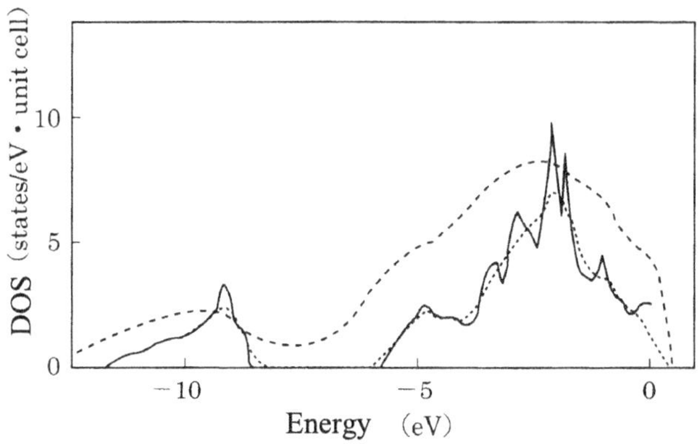

Fig. 4.17. Density of states for ferromagnetic MnSb (*solid curve*) and XPS spectrum (*dashed curve*). The *dotted curve* shows DOS broadened with a Gaussian of full with at half height = 0.6 eV (Reproduced with permission from [4], Copyright (2009) by the American Physical Society)

volume decrease. In the temperature region above T_C, reverse change to the NiAs-type structure takes place at $T_t = 398$ K (see Fig. 3.3. The NiAs-type and MnP-type structures are also shown in Fig. 4.38). In the temperature region, $T > T_t$, paramagnetic susceptibility obeys approximately the Curie–Weiss law. According to observation carried out by Ido et al., the χ^{-1}-T curve shows a slight upward convex in this temperature range. Magnitude of magnetic moment evaluated from the χ^{-1}-T curve in the high-temperature region is 3.8 μ_B. This is larger than saturate moment in the ferromagnetic state 3.4 μ_B.

In the intermediate temperature range ($T_C \leq T \leq T_t$) where the crystal structure is of the MnP-type, the χ^{-1}-T curve shows anomalous behavior that the curve shows a broad peak, and crystal volume increases rapidly with increasing temperature (see Figs. 3.2 and 3.3, etc.). Crystal volume at T_t is almost equal to the volume just below T_C. Such anomalous behavior of susceptibility and thermal expansion in the MnP-type phase are more remarkable in $MnAs_{1-x}P_x$ ($0 \leq x \leq 0.275$). $CrSb_{1-x}As_x$ and $Cr_{0.4}Mn_{0.6}As$ in the MnP-type phase also show such anomalous behavior as shown in Part I. According to neutron diffraction measurements by Schwartz [22], magnetic moment of MnAs and $MnAs_{1-x}P_x$ in the MnP-type phase is smaller than the moment in the NiAs-type phase. They also found that magnitude of magnetic moment shows smooth change with temperature.

According to measurements of magnetization process carried out by the group of Osaka University (Fig. 3.4 and Fig. 3.15–3.17, etc.), magnetic moment of $MnAs_{1-x}P_x$ depends little on magnetic fields at temperatures below T_C, whereas it depends considerably on magnetic fields at around T_t. In the MnP-type phase in the temperature region just above T_C, field-induced metamagnetic magnetization curve with hysteresis was observed. As shown in Fig. 3.4, temperature dependences of magnetization measured in 20 and 400 kOe revealed that magnetic moment in 20 kOe shows a discontinuous change at T_C. This change is considered due to the structural phase transition at T_C. On the other hand, magnetization in 400 kOe continuously changes with temperature, which suggests that MnAs remains to be of the NiAs-type structure even in the temperature region above T_C. These results mentioned earlier indicate that the metamagnetic behavior is accompanied with the structural transition from the MnP-type phase (low-fields) to the NiAs-type phase (high-fields).

Yamaguchi et al. carried out neutron diffraction measurements for the ferromagnetic state of MnAs [3]. They reported that a magnetic moment $0.23\mu_B$ exists at an As-site and it is antiparallel to magnetic moment at Mn-site. Their neutron diffraction measurements also revealed that an Sb atom in MnSb has moment $0.3\mu_B$ in the opposite direction of the moment of Mn atom.

4.3.2 Spin Fluctuation and Magnetism

In the itinerant model, correlation between electrons is a very important factor to discuss magnetic properties in finite temperature. Moriya and his co-workers developed the theory in which the correlation is treated in terms of fluctuation of spin density. They succeeded in the unified theory in describing both the extreme case of weak ferromagnetism, where the spin fluctuation has a small amplitude and local character in the $\boldsymbol{q}$-space, and another extreme the case of local moment, where all $\boldsymbol{q}$-modes of spin fluctuation and their mode–mode coupling are important [23–25].

Motizuki and Katoh investigated magnetism of MnAs [26, 27] by applying the spin-fluctuation theory within the formalism of Moriya–Takahashi [24] and Usami–Moriya [25].

We start from the following single-band Hubbard Hamiltonian:

$$\begin{aligned}\mathscr{H} &= \mathscr{H}_0 + \mathscr{H}_1 \\ &= \sum_{\sigma,j,\ell} t_{j\ell}\, a^{\dagger}_{j\sigma} a_{\ell\sigma} + \sum_{j}\left[\frac{U}{4}n_j^2 - JS_j^2\right],\end{aligned} \tag{4.3}$$

where the symbol $t_{j\ell}$ denotes a transfer integral, $a^{\dagger}_{j\sigma}$ a creation operator which creates an electron with spin σ at a site j. n_j and S_j in $\mathscr{H}_1$ are operators of charge density and spin density, respectively. U and J are an effective intra-atomic Coulomb integral and an effective intra-atomic exchange integral, respectively. Assuming that five d orbitals degenerate, U and J are related to U' and J' (Coulomb and exchange integrals for a single orbital Hubbard Hamiltonian, respectively) as

$$U = \frac{1}{5}(9U' - 4J')$$

$$J = \frac{1}{15}(U' + 4J').$$

By using the Stratonovich–Hubbard transformation, we can transform this many-body problem into a single-particle problem under a magnetic field (ξ) and an electric field (η) fluctuating temporally and spatially. Within the static approximation, a partition function Z (or free energy F) can be written in the functional integral form as

$$Z = \mathrm{e}^{-\beta F} = \mathrm{Tr}\left[\mathrm{e}^{-\beta(\mathscr{H}_0 - \mu\hat{N})}\right]\mathrm{e}^{-\beta\Delta F} \tag{4.4}$$

and

$$\mathrm{e}^{-\beta\Delta F} = \int \prod_j \mathrm{d}\xi_j \mathrm{d}\eta_j \exp(-\beta\Psi[\xi,\eta]). \tag{4.5}$$

Here, μ is the chemical potential, $\hat{N}$ a total electron number operator, and $\beta = 1/k_{\mathrm{B}}T$. The functional $\Psi[\xi,\eta] = \Psi_0[\xi,\eta] + \Psi_1[\xi,\eta]$, is given by

$$\Psi_0[\xi,\eta] = \frac{\pi}{\beta}\sum_j\left[\xi_j^2 + \eta_j^2\right] \tag{4.6}$$

and

$$\mathrm{e}^{-\beta\Psi_1[\xi,\eta]} = \left\langle T_\tau \exp\left[-\int_0^\beta \mathrm{d}\tau \sum_j\left\{\bar{c}_1\boldsymbol{\xi}_j\cdot\boldsymbol{S}_j(\tau) + \bar{c}_2\eta_j n_j(\tau)\right\}\right]\right\rangle, \tag{4.7}$$

where

$$\bar{c}_1 = \sqrt{4\pi J/\beta}, \qquad \bar{c}_2 = \sqrt{-\pi U/\beta}, \tag{4.8}$$

and ξ_j and η_j denote local magnetic and electric fields, respectively, operating at the jth site. In addition, we introduce a quantity

$$x_\alpha = \frac{1}{N_0\beta}\sum_q{}' \xi_{q\alpha}\xi_{-q\alpha} \qquad (\alpha = x,y,z), \tag{4.9}$$

where ξ_q is a Fourier component of the fluctuating magnetic field ξ_j and N_0 is the number of lattice points in the system. We note that Σ_q' means that the term $\boldsymbol{q} = 0$ is excluded. x_α in (4.9) is a square average of local amplitude of the ξ-field, and related to a square average of local amplitude of spin fluctuation as

$$\left\langle (S_{j\alpha} - \langle S_{j\alpha}\rangle)^2\right\rangle = \frac{\pi}{J}\left(x_\alpha - \frac{1}{2\pi\beta}\right). \tag{4.10}$$

By assuming the mode–mode coupling of spin fluctuation to be local and applying a saddle-point approximation for the charge field, Moriya and the co-workers introduced the following form of the functional:

$$\Psi_1[\xi] = -\frac{2\pi J}{\beta}\sum_q{}'\sum_\alpha X_{q\alpha}[x,\xi_0]\xi_{q\alpha}\xi_{-q\alpha} + N_0 L[x,\xi_0]. \tag{4.11}$$

The first term of (4.11) represents the nonlocal part, where terms independent on relative direction of spin fluctuation are neglected. An appropriate function of $\boldsymbol{q}$ is assumed for $X_{q\alpha}[x, \xi_0]$. The second term of (4.11) represents the local part, for which Usami and Moriya obtained an expression for $L[x, \xi_0]$ by using the coherent potential approximation (CPA).

From the functional Ψ_1 thus obtained, the partition function of the system is calculated and then various quantities such as paramagnetic susceptibility and local magnetic moment can be evaluated successfully.

We show here the results of calculations carried out for paramagnetic MnAs and $MnAs_{1-x}P_x$ [26, 27]. First, we simplified the density of states calculated by the self-consistent APW method by replacing the curves with a set of straight lines and then obtained model density of states as shown in the lower part of Fig. 4.18. We use the model density of states for the present calculations. For example, the curve denoted by (1) in Figs. 4.18 and 4.19 show calculated local magnetic moment $\sqrt{\langle S_j^2 \rangle}$ and inverse paramagnetic susceptibility χ^{-1}, respectively. They are calculated with the model density of states (1) and in the case that electron number per orbital n is 0.65 and U and J are equal to 2.8 (in unit of 5/6 eV). As shown in Fig. 4.19 the calculated χ^{-1}-T curves show the Curie–Weiss behavior with weak upward convex. Magnetic moment estimated from the χ^{-1}-T curve is 2.58μ_B, which is slightly larger than the value 2.30 μ_B evaluated from the model density of state and the Hartree–Fock approximation at 0 K. As seen from Figs. 4.19 and 4.20, there is thus good correspondence between the present calculations and the experimental results except for the numerical values. For example, Curie temperature estimated from an equation $\chi^{-1} = 0$ is only 1/10 of the observed value. It is needed to point out that the relation

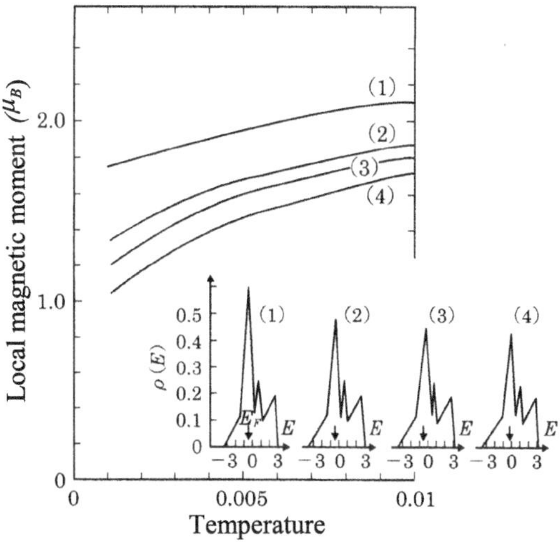

Fig. 4.18. Calculated local magnetic moments m_{loc} using the model density of states (1)–(4). The model density of states (1) corresponds to the NiAs-type MnAs, (2)–(4) correspond to MnAs distorted to the MnP-type structure (Reproduced with permission from [26])

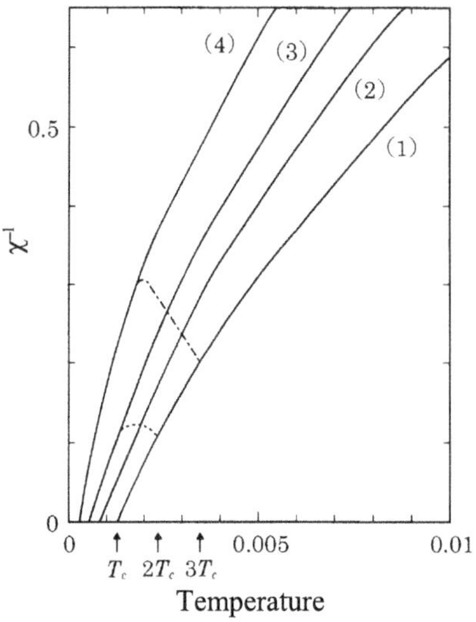

Fig. 4.19. Inverse paramagnetic susceptibility calculated for the model densities of states (1)–(4). The *dotted* and *dash-dotted curves* represent χ^{-1} in the intermediate temperature range $T_C \leq T \leq T_t$ with $T_t = 2T_C$ and $T_t = 3T_C$, respectively (Reproduced with permission from [26])

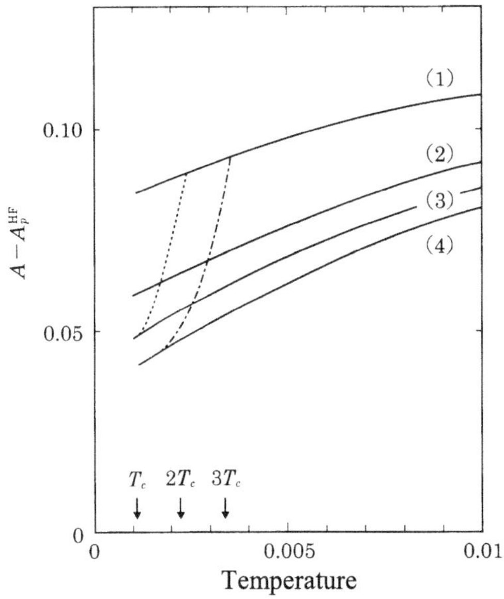

Fig. 4.20. Temperature dependence of $(A_p - A_p^{HF})$ calculated for the model densities of states (1)–(4) shown in Fig. 4.18. The *dotted* and *dash-dotted curves* represent $(A_p - A_p^{HF})$ in the intermediate temperature range $T_C \leq T \leq T_t$ with $T_t = 2T_C$ and $T_t = 3T_C$, respectively (Reproduced with permission from [27])

$\chi^{-1} = 0$ gives T_C for the case of a second-order transition and the transition at T_C of MnAs is of a first order. We may expect a better estimation of T_C by considering effect of degeneracy of orbitals and quantum effects which are neglected in the spin-fluctuation theory.

Next, we investigate paramagnetic susceptibility of MnAs in the intermediate temperature range of $T_C \leq T \leq T_t$ where the crystal structure is of the MnP-type. As noted in Sect. 4.2, the APW band calculations have revealed that the height of the sharp peak of density of state near the Fermi level in the NiAs-type structure becomes smaller with lattice distortion to the MnP-type structure. In order to take such change of band structure into account, we modified the model density of states. The model density of states in the MnP-type structure is shown in the lower part of Fig. 4.18 for the three cases denoted by (2), (3), and (4), which correspond to the cases where $\rho(E_F)$ is 90%, 85%, and 80% of $\rho(E_F)$ of (1) in the NiAs-type phase, respectively. The solid curves denoted by (2), (3), and (4) in Figs. 4.18 and 4.19 display the temperature variations of the local magnetic moment and inverse paramagnetic susceptibility calculated for the model densities of states (2), (3), and (4), respectively. It is seen from Figs. 4.18 and 4.19 that the local magnetic moment decreases in the order (1), (2), (3), and (4), resulting in steeper inclination of the χ^{-1}-T curves. In order to explain the observed χ^{-1}-T curve of MnAs, in Fig. 4.19 we start from a point on the curve (1) in the high-temperature region and consider the χ^{-1}-T curve in the process of lowering temperature. With decreasing temperature, χ^{-1} decreases first along the curve (1), and in the MnP-phase below the transition temperature T_t, χ^{-1} changes to the values of points on the curves (2), (3), and (4) successively because the lattice distortion increases as temperature decreases. The dotted- and the dashed-curves show temperature dependences of χ^{-1} for two cases of (i) $T_t = 2T_C$ and ii) $T_t = 3T_C$, respectively. For the case (i), we assumed that the lattice distortion changes monotonically with temperature in the temperature range between T_t and T_C. For the case (ii), the monotonical change of distortion in the temperature region between T_t and 1.5 T_C, and constant distortion in the temperature region between 1.5 T_C and T_C are assumed. Measurements of lattice distortion for MnAs and $MnAs_{0.8}P_{0.2}$ has revealed that $MnAs_{0.8}P_{0.2}$ takes the MnP-type structure in wider temperature range than MnAs (see, for example, Fig. 3.19 and [28]). It also has been found that temperature dependence of the lattice distortion of $MnAs_{0.8}P_{0.2}$ is large only in a narrow temperature region just below T_t, but almost constant in other temperature regions below T_t [28]. Thus, the χ^{-1}-T curve calculated for the case (i) corresponds to MnAs and that for the case (ii) to $MnAs_{0.8}P_{0.2}$. The χ^{-1}-T curves of these compounds show a peak in the temperature region of the MnP-type phase. The peak of $MnAs_{0.8}P_{0.2}$ is especially remarkable (see, for example, Figs. 3.2 and 3.13).

As mentioned earlier, we have shown that the anomalous temperature dependences of susceptibility observed for MnAs and $MnAs_{1-x}P_x$ compounds are reasonably understood by the spin-fluctuation theory including the effect of lattice distortion of the MnP-type structure.

4.3.3 Spin Fluctuation and Elastic Properties

We describe in this section how to explain anomalous elastic properties of MnAs and $MnAs_{1-x}P_x$ in the MnP-type phase from the viewpoint of itinerant electrons taking the effect of spin fluctuation into account [27].

Cohesive pressure of electron system is given by $P = -\frac{\partial F}{\partial V}$ with the free energy F. For simplicity we assume that electronic energy ε_k is proportional to band width W and that only W depends on volume. The pressure P in this approximation is written as

$$P = \lambda A/V. \tag{4.12}$$

In this equation, λ is defined by $\lambda = -\frac{V}{W}\frac{\partial W}{\partial V}$ and A is given by

$$A = \sum_k \varepsilon_k \frac{\partial F}{\partial \varepsilon_k}. \tag{4.13}$$

From free energy evaluated by using the spin-fluctuation theory described in Sect. 4.3.2, we calculated the quantity A for paramagnetic NiAs-phase and paramagnetic MnP-phase. The free energy per atomic site is written as a summation of F_0 and ΔF, which are terms irrelevant and relevant, respectively, to the spin fluctuation. The quantities F_0 and ΔF are given by

$$F_0 = -\frac{1}{N_0\beta}\sum_{k\sigma} \ln\left[1 + e^{\beta(\mu-\varepsilon_k)}\right] \tag{4.14}$$

and

$$\Delta F = \pi\left(x + \frac{|\eta_j|^2}{\beta}\right) + L(x), \tag{4.15}$$

where the nonlocal term has been omitted. By using an expression of the local term $L(x)$ obtained within the coherent potential approximation (CPA), we can calculated the quantity A from (4.13). The parameter A_p, which means A of the paramagnetic state, is given by

$$A_\mathrm{p} = -\frac{2}{\pi}\int d\varepsilon f(\varepsilon + \mu) \times \mathrm{Im}\left\{\left(\varepsilon + \mu - \Sigma_\mathrm{p} - \gamma_\mathrm{p}\right) F_\mathrm{p}\right\}, \tag{4.16}$$

where f is the Fermi–Dirac distribution function and ε energy of electrons, and other symbols in (4.16) are explained next. We note that $|\eta_j|^2$ and x in the expression of ΔF are assumed not to depend on ε_k explicitly. The chemical potential μ is determined by

$$-\frac{2}{\pi}\int d\varepsilon f(\varepsilon + \mu)\,\mathrm{Im}F_\mathrm{p} = n, \tag{4.17}$$

where n is electron number. The symbol γ_p is $nU/2$, F_p is a diagonal element of single-particle Green's function, and Σ_p denotes site-independent self-energy. Within the Hartree–Fock approximation in which the spin fluctuation is neglected, the local magnetic moment is zero and $\Sigma_p = 0$. In this case, A_p becomes A_p^{HF} given by

$$A_p^{HF} = 2\int d\varepsilon f(\varepsilon + \gamma_p)\,\varepsilon\,\rho(\varepsilon), \tag{4.18}$$

where $\rho(\varepsilon)$ is the density of states. Contribution of the spin fluctuation to A is given by $(A_p - A_p^{HF})$. Change of pressure ΔP is related by the compressibility κ to change of volume ΔV as

$$\frac{\Delta V}{V} = \kappa \Delta P \tag{4.19}$$

We see from (4.12) that ΔP is written by ΔA, therefore volume change in (4.19) is given by

$$\begin{aligned}\frac{\Delta V}{V} &= \lambda\,\kappa\Delta(A_p - A_p^{HF})/V + \lambda\,\kappa\Delta A_p^{HF}/V \\ &\equiv \frac{\Delta V_1}{V} + \frac{\Delta V_2}{V}.\end{aligned} \tag{4.20}$$

The first and the second terms denote volume changes arising from the spin fluctuation and that from change of electronic energy, respectively.

Calculated $(A_p - A_p^{HF})$ is plotted against temperature in Fig. 4.20. The curves (1)–(4) are results calculated for the model densities of states (1)–(4) shown in Fig. 4.18, respectively. Since $(A_p - A_p^{HF})$ arises from the spin fluctuation, temperature dependence of this quantity reflects temperature dependence of local amplitude of spin fluctuation plotted in Fig. 4.18. As seen in Fig. 4.20 that $(A_p - A_p^{HF})$ at each temperature decreases in the order (1), (2), (3), and (4), which means that spin fluctuation is suppressed as the lattice distortion to the MnP-type increases. To explain anomalous thermal expansion observed for MnAs, etc., (see Figs. 3.3, 3.19, etc., in Part I) we consider temperature dependence of $(A_p - A_p^{HF})$ in the process of decreasing temperature; $(A_p - A_p^{HF})$ decreases first along the curve (1), and in the temperature region of $T_C \leq T \leq T_t$ below T_t, it changes successively to the points on the curves (1) →(2) →(3) →(4) as the lattice distortion increases. The dotted- and dashed-dotted-curves denote temperature dependence of $(A_p - A_p^{HF})$ for the case $T_t = T_C$ and $T_t = 3T_C$, respectively. These results clearly indicate that the spin fluctuation plays an important role on the remarkable volume change with temperature for MnP-type MnAs and $MnAs_{1-x}P_x$, because temperature dependence of $(A_p - A_p^{HF})$ gives rise to the volume change.

Next, we investigate the volume change at T_C. Using values $\kappa = 4.5 \times 10^{-11}$ m^2/N, $V = 34 \times 10^{-30}$ m^3, and $\lambda = 5/3$ (estimated by Heine [29]), we evaluated at T_C the quantity $\Delta V_1/V$ from $(A_p - A_p^{HF})$ to be 3.4%, 4.8%, and 5.8% for the changes of $(A_p - A_p^{HF})$ between curves (1) and (2), (1) and (3), and (1) and (4), respectively. We also estimated $\Delta V_2/V$ comparable to $\Delta V_1/V$ from the tight-binding calculation for the MnP-type structure, which means that decrease of local magnetic moment due to the structural transition and energy change of electron system have comparable contributions to the discontinuous volume change at T_C in MnAs (see Fig. 3.3 in Part I). Calculated values of $\Delta V/V$ is 7.1% for the change (1) →(2), 9.5% for (1) →(3), and 12% for (1) →(4), respectively. Experimental volume change at T_C is 2% for MnAs and 8 ∼ 10% for $MnAs_{1-x}P_x$.

Finally, we note that we estimated $\lambda \simeq 1.1$ from our band calculation for MnAs. If we use this value of λ instead of Heine's result $\lambda = 5/3$, the calculated values of $\Delta V/V$ become smaller by about 2/3 than the value mentioned earlier.

4.3.4 Paramagnetic Susceptibility of CoAs and FeAs

The compounds FeAs and NiAs crystallize in the MnP-type and the NiAs-type structures, respectively, while CoAs shows the structural phase transition at 1,250 K from the MnP-type to the NiAs-type structures as temperature rises. Below this temperature, the crystal structure of CoAs becomes the MnP-type. FeAs takes the MnP-type structure at all temperatures, and NiAs remains in the NiAs-type structure except at very temperatures.

Morifuji and Motizuki calculated paramagnetic susceptibility χ of these compounds by using the spin-fluctuation theory and nonmagnetic bands calculated by the APW method [30]. It has been revealed that χ of NiAs is almost temperature-independent, while χ's of CoAs and FeAs, which have the MnP-type structure, shows anomalous temperature dependence. According to the experimental data [31], inverse susceptibilities of CoAs and FeAs obey the Curie–Weiss law in high temperature region and increase for further decreasing of temperature.

The temperature dependences of paramagnetic susceptibility of CoAs and FeAs are similar to that of a weak ferromagnetic semiconductor FeSi. Takahashi and Moriya explained anomalous temperature dependence of susceptibility of FeSi on the basis of the spin-fluctuation theory and characteristic shape of density of states [32]. According to their theory, negative mode–mode coupling is thermally induced when the Fermi level lies at a dip of density of state and the density of states just below the Fermi level has a large value. In this situation, the negative mode–mode coupling gives rise to a dip of the χ^{-1}-T curve and a rapid increase of χ^{-1} in the temperature region below the dip. Furthermore, the Curie–Weiss behavior of susceptibility occurs in the temperature region above temperature of the dip. The Curie–Weiss behavior is brought about by the saturation of local amplitude of spin fluctuation.

Morifuji and Motizuki calculated χ^{-1}-T curves for CoAs and FeAs by taking account of the spin-fluctuation effect, as well as the fact that the densities of states for FeAs and CoAs shown in Fig. 4.9 are similar to that of FeSi [33, 34].

We show in Fig. 4.21 the calculated results for CoAs. We used the model density of states, shown in the inset, in which the calculated curve is replaced by fragments of straight lines. We see that the temperature dependence of inverse susceptibility χ^{-1} agrees qualitatively with that of measurement plotted by × [31]. This result indicates that the anomalous temperature dependence of paramagnetic susceptibility χ^{-1} of

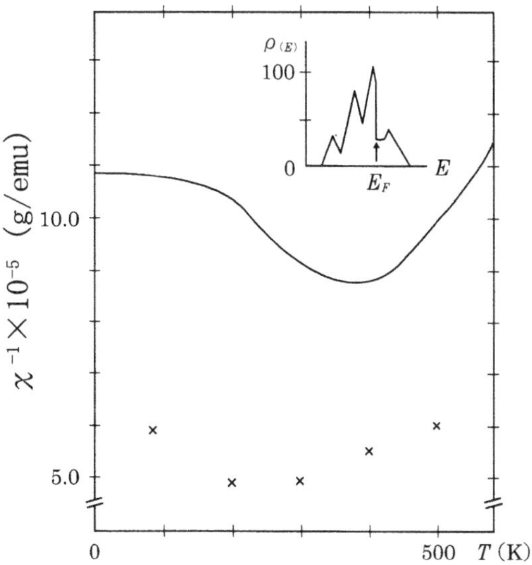

Fig. 4.21. Inverse paramagnetic susceptibility of CoAs calculated by using the model density of states inserted in the figure is plotted by the solid curve. The observed susceptibilities at several temperatures are shown by ×

CoAs is well understood in terms of the spin-fluctuation effect except discrepancy in magnitude of χ^{-1} and the temperature of the dip. We have obtained similar results for FeAs.

4.4 Fermi Surface of NiAs and the de Haas–van Alfen Effect

We can investigate Fermi surface by observing the de Haas–van Alfen effect. Since the cross section normal to an applied magnetic field is related to period of the de Haas–van Alfen oscillation, it is possible to investigate shape of Fermi surfaces from the de Haas–van Alfen measurements carried out in various directions of applied magnetic fields.

Recently, various techniques and theories for band calculation have also been developed to calculate more accurate band structures. As for the correlation between electrons, the local density approximation, which is usually used, may be insufficient because average electron density is used. Recently, improvement of the LDA such as GGA has been carried out. It is thus important to justify calculated bands by examining the shape of Fermi surfaces.

The Group of Kamimura measured the de Haas–van Alfen oscillation using single crystals of NiAs and NiSb with the NiAs-type structure [35]. Since the NiAs-type structure has two metal atoms and two anions in the unit cell, the dispersion curves near the Fermi level have a complicated dependence on wavevector due to mixing between d-orbitals of Ni and p-orbitals of As or Sb. Many branches are observed in the experiments, and therefore it is difficult to reproduce the shape of Fermi surfaces from de Haas–van Alfen measurements.

Recently, however, Harima developed a computer program to calculate de Haas–van Alfen frequencies from Fermi surfaces obtained by band calculations [36]. It has enabled us to easily make comparison between calculated band structures and de Haas–van Alfen measurements. To make direct comparison, we first constructed Fermi surfaces of NiAs from results of band calculation and determined dependences of the de Haas–van Alfen frequencies in the direction of magnetic field, and then compared them with the experimental data [35, 37].

We calculated initially band structure of nonmagnetic NiAs and NiSb by using the APW method, and later the FLAPW method to obtain more accurate band structures. We applied the LDA with the GL formula for the exchange and correlation interaction. Scalar relativistic effects were taken into account and the spin–orbit interaction was omitted in the calculations. The program codes TSPACE and KANSAI-92 were used for these calculations.

Dispersion curves of NiAs and NiSb are shown in Fig. 4.22a, b, respectively. The bands in the low-energy region consist of As-4s or Sb-5s orbitals. The 16 bands above the gap are p–d mixing bands consisting of Ni-3d and As-4p (or Sb-5p)

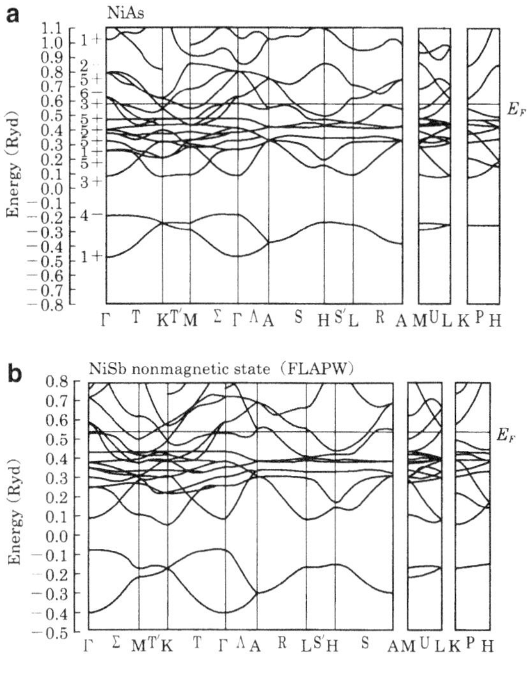

Fig. 4.22. Dispersion curves of nonmagnetic band of (**a**) NiAs and (**b**) NiSb (Fig. 4.22a is reproduced form [35])

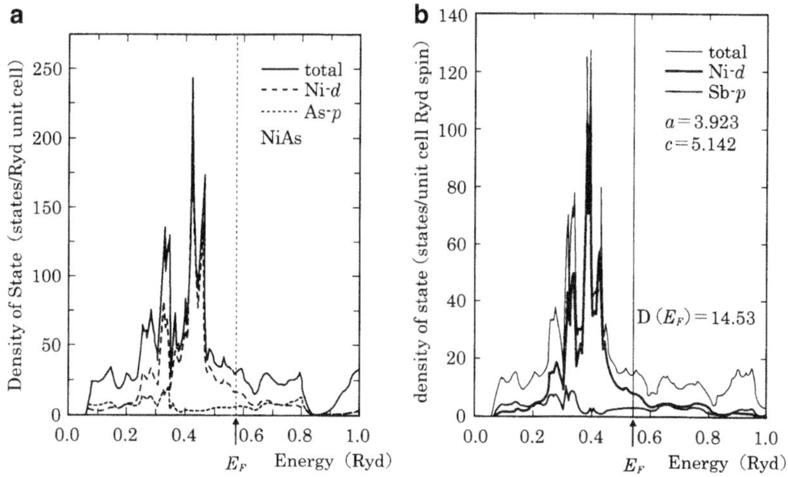

Fig. 4.23. Density of states of nonmagnetic band of (**a**) NiAs and (**b**) NiSb (Fig. 4.23a is reproduced form [35])

orbitals. We show the total densities of states of the *p–d* mixing bands $\rho(E)$ with the components of Ni-*d* orbitals and As- (or Sb-) *p* orbitals for NiAs and NiSb in Fig. 4.23a, b, respectively. The density of states in both compounds can be divided into three main parts: *p–d* bonding band, nonbonding band consisting mainly of *d* orbitals in the middle region of energy, and *p–d* antibonding band in the highest region of energy. The gross feature of the density of states is similar to that of MnAs. Total width of the *p–d* mixing band is larger than that of MnAs shown in Figs. 4.2 and 4.3. The Fermi level is an energy far above the energy region with the large peak of the density of states, $\rho(E_F) = 29.12$ (states/Ry unit cell) for NiAs and 14.53 (states/Ry unit cell) for NiSb are obtained. These values are much smaller than 156 (states/Ry unit cell) of MnAs, which is consistent with the fact that MnAs is a ferromagnet but NiAs and NiSb are paramagnetic.

The value of $\rho(E_F)$ of NiAs evaluated from measurement of temperature dependence of specific heat is 34.54 (states/Ry unit cell) [37], which is somewhat larger than the calculated value of 29.12, however, the difference between experimental and calculated values in NiAs is smaller than that in MnAs. This fact suggests that electron–electron interaction and electron–phonon interaction in NiAs is weaker than those in MnAs. Width of the density of states below the Fermi level for both NiAs and NiSb is about 6 eV ($\simeq$0.44 Ry) and there is a sharp peak at 2 eV ($\simeq$0.15 Ry) energy lower than the Fermi level. Such character of the density of states corresponds well to the results of photoemission measurements [38].

Figure 4.24 shows calculated Fermi surfaces of NiAs. NiAs has three Fermi surfaces: two hole surfaces with cylindrical shape around the ΓA-axis and an electron surface whose shape is somewhat complicated.

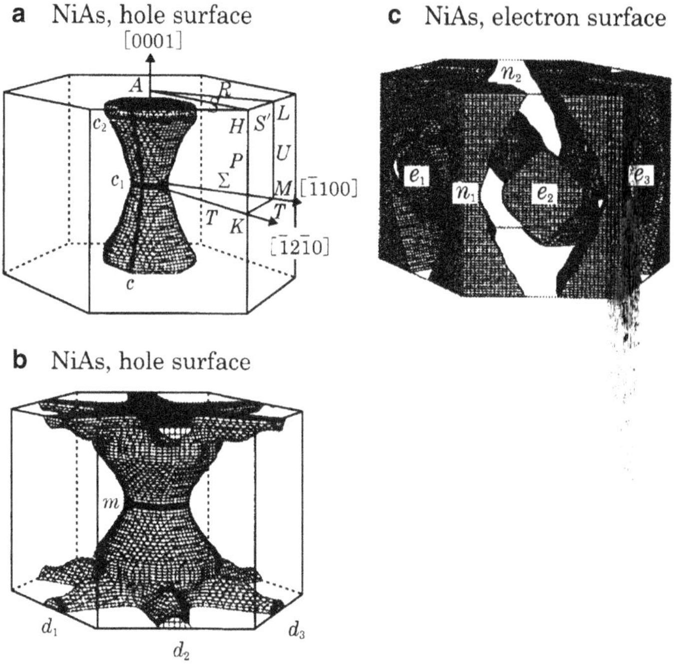

Fig. 4.24. Fermi surfaces of nonmagnetic state of NiAs (Reproduced form [35])

The de Haas–van Alphen frequencies calculated from the Fermi surfaces are plotted as a function of angle of direction of magnetic field in Fig. 4.25. The branches labeled *e*, *m*, and *d* in Fig. 4.25 seem to correspond generally well to the experimental branches labeled ε, μ, and δ as shown in Fig. 4.26 [35].

However, the α and β branches in Fig. 4.26 cannot be explained by the present calculations, and the *c*- and *n*-branches which are seen in the theoretical curves are not measured. The absence of observations of the *c*- and *n*-branches may be explained by a possible structural transformation due to nesting effect between the hole surface around the ΓA-axis and the electron surface around the KH-axis. Recent X-ray diffraction measurements have revealed that NiAs shows a structural transformation at a very low temperature [39]. Due to this structural transformation, it is considered that the hole surface around the ΓA-axis and the electron surface around the KH-axis disappear.

Figures 4.27 and 4.28 show the Fermi surfaces and the calculated de Haas–van Alfen frequency of NiSb, respectively [37, 40]. The results of measurements are shown in Fig. 4.29. As seen in NiAs, there is good correspondence between the *e* branch in Fig. 4.28 and the ε branch in Fig. 4.29. Unlike the case of NiAs, the α- and the β-branches which correspond to the c_1- and the n_1-branches in Fig. 4.28 are observed in NiSb. This result corresponds well to the fact that NiSb does not show structural transformation, but remains in the NiAs-type structure even in low temperature.

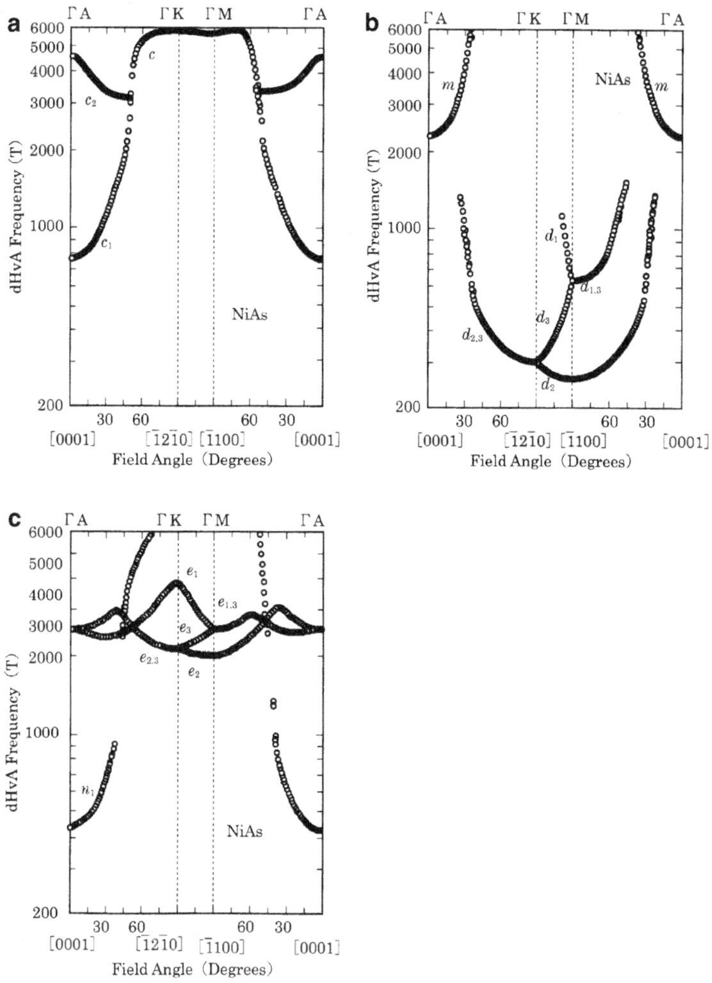

Fig. 4.25. Calculated de Haas–van Alfen frequencies of nonmagnetic NiAs (Reproduced form [35])

4.5 Pressure Effect on Magnetic State of CrTe, CrSe, and CrS

CrTe becomes a ferromagnet below $T_C = 340\,\mathrm{K}$. By application of high pressure, T_C decreases remarkably and the magnetic moment also decreases, which suggests a kind of pressure-induced phase transition.

To clarify the effects of high pressure on magnetic properties of CrTe, we carried out band calculations for nonmagnetic, ferromagnetic, and antiferromagnetic states for various values of the unit cell volume. In these calculations, the axial ratio c/a is fixed to the value at room temperature. For the antiferromagnetic calculation, the sublattice magnetic moments are assumed to be in the c-plane and are antiparallel

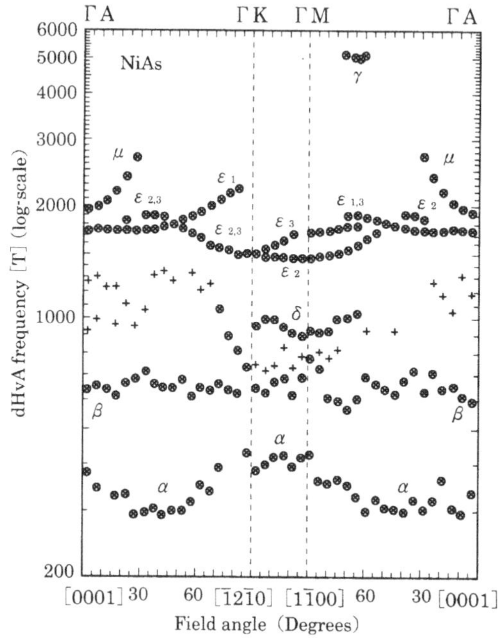

Fig. 4.26. Experimental de Haas–van Alfen frequencies of NiAs (Reproduced form [35])

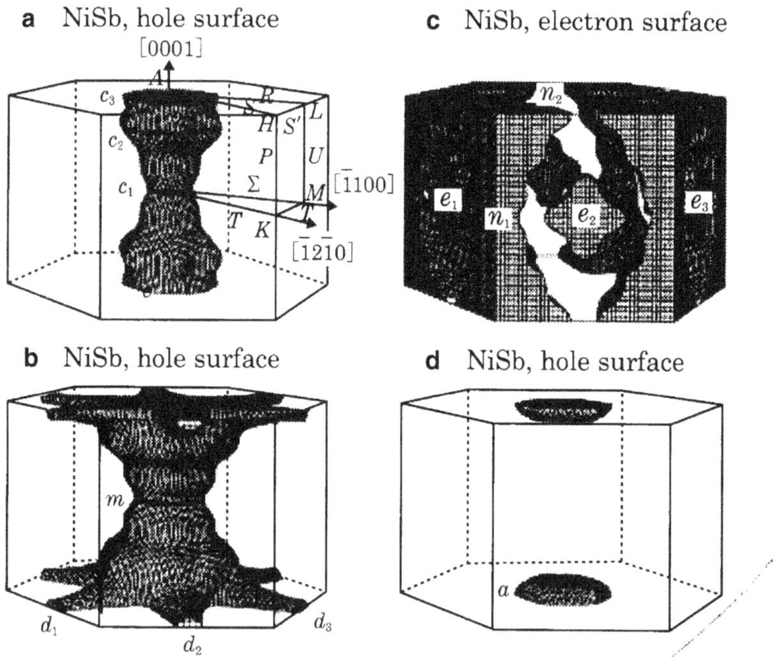

Fig. 4.27. Fermi surfaces of nonmagnetic NiSb

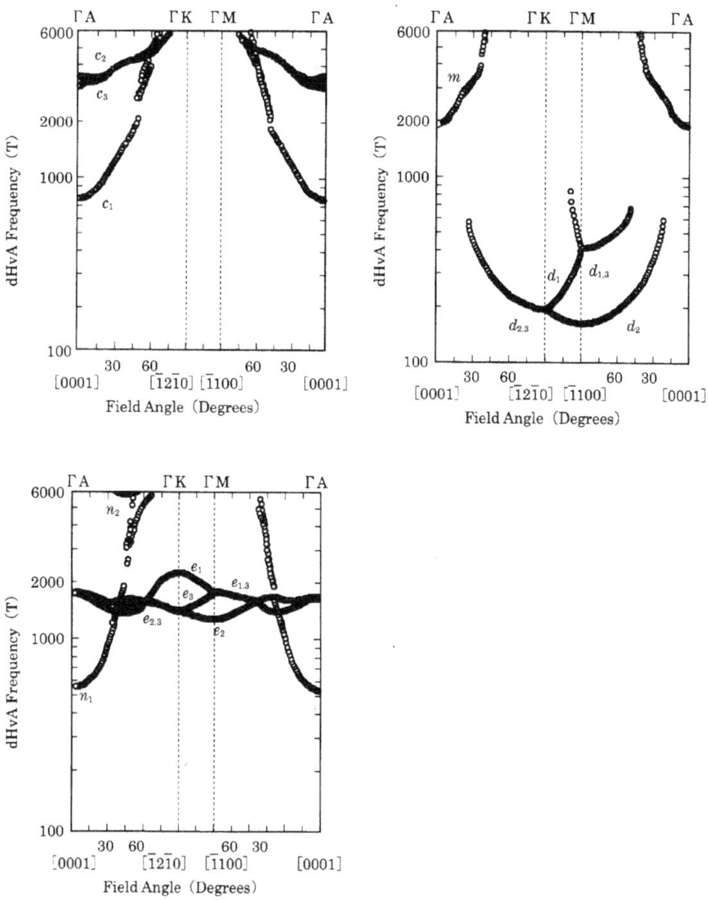

Fig. 4.28. Calculated de Haas–van Alfen frequencies of NiSb

between adjacent c-planes. We employed the APW method for the first calculations, later we used the FLAPW method to obtain more accurate band structures [41–43]. In this section, we show the results obtained by the FLAPW method. The density of states of ferromagnetic CrTe calculated by the FLAPW method is shown in Fig. 4.14.

In Fig. 4.30, we show the calculated total energy as a function of the lattice constant a for nonmagnetic, ferromagnetic, and antiferromagnetic states in CrTe [41]. The total energy for the ferromagnetic state takes the minimum value at $a = 3.970$ Å, which agrees well with the experimental value $a = 3.981$ Å. We can express the total energy curve of ferromagnetic state by a function

$$E_{\text{Total}} = E_0 + b_2(a - a_0)^2 + b_3(a - a_0)^3 + b_4(a - a_0)^4, \tag{4.21}$$

where $E_0 = -31357.17$ (Ry/unit cell), $a_0 = 3.970$, $b_2 = 0.471$, $b_3 = -0.575$, and $b_4 = 1.686$.

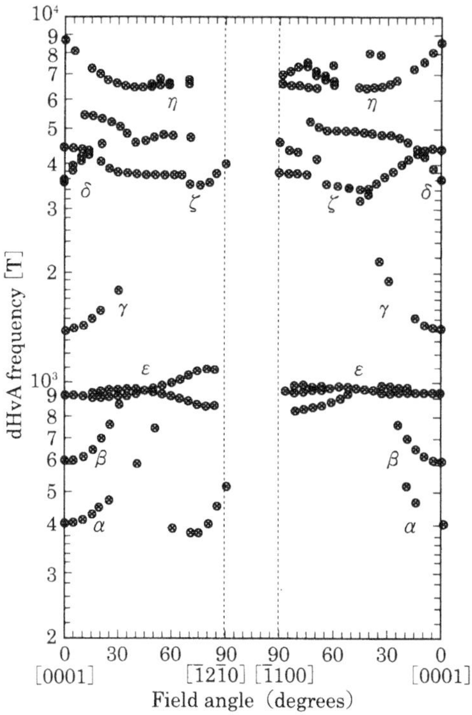

Fig. 4.29. Experimental de Haas–van Alfen frequencies observed for NiSb

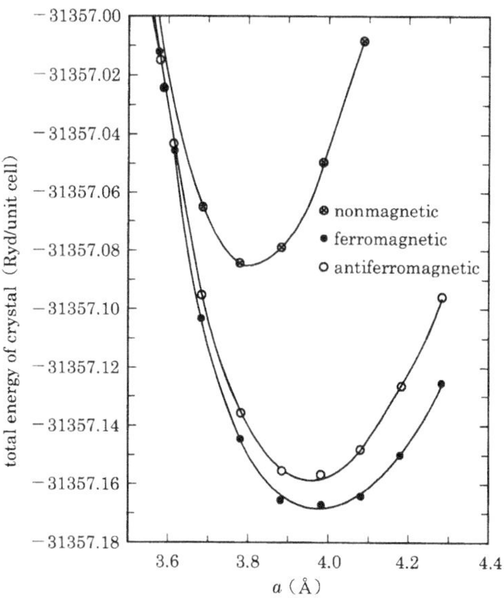

Fig. 4.30. Total energy curves for ferromagnetic, antiferromagnetic, and nonmagnetic CrTe as a function of lattice constant a (Reproduced from [41])

Since the APW method gives the minimum energy at the lattice constant $a = 4.180$ Å, we see that the FLAPW method brings about improvement of calculated results. Generally, a full potential calculation reproduces more correctly the experimental values of lattice constant than the APW method. As shown in Fig. 4.30, the ferromagnetic state is most stable at $a = 3.981$ Å, which is consistent with the fact that CrTe is a ferromagnet under atmospheric pressure.

It is seen in Fig. 4.30 that the antiferromagneic state has smaller energy than that of the ferromagnetic state for the lattice constant smaller than $a = 3.58$ Å, which indicates that pressure-induced phase transition from the ferromagnetic state to the antiferromagnetic state can occur. The inset of Fig. 4.31 shows relation between calculated electronic pressure and the lattice constant a. From this curve, we see that the pressure-induced magnetic transition may occur at a critical pressure of about 40 GP which corresponds to the lattice constant 3.58 Å. The APW calculation gives critical pressure 20 GP. In Fig. 4.31 total magnetic moment of Cr and Te in the ferromagnetic CrTe are plotted against the lattice constant a. Large part of the magnetic moment arises from Cr atom, however, Te atom has a small negative magnetic moment in the opposite direction. (see also Sect. 4.2.4).

The total magnetic moment of CrTe is calculated to be $3.1\mu_B$/formula at $a = 3.970$ Å. This value is not so different from observed moment $2.29\mu_B$/formula. The APW calculation gives magnetic moment $3.9\mu_B$/formula.

We can evaluate electron pressure P from the equation $P = -\mathrm{d}E_{\mathrm{Total}}/\mathrm{d}V$. Since the unit cell volume is given by $V = 1.35a^3$, the pressure P is expressed as $P = -0.247a^{-2}\,\mathrm{d}E_{\mathrm{Total}}/\mathrm{d}a$. The bulk modulus, which is given by $B = -\mathrm{d}P/\mathrm{d}\log V$ ($V = V_{\mathrm{eq}}$), is calculated to be $B = 404$ kbar, or the compressibility (B^{-1}) 2.43

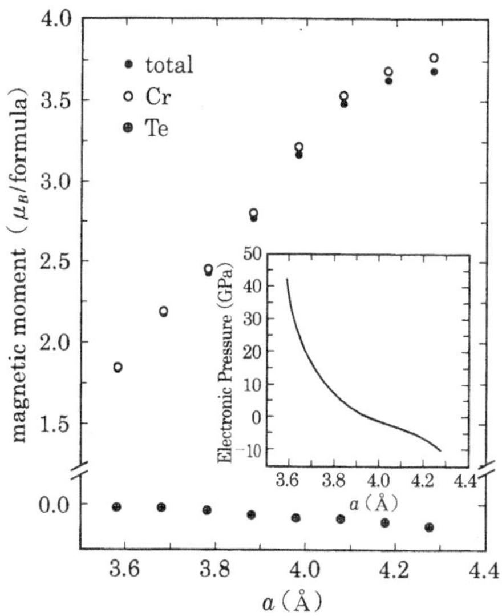

Fig. 4.31. Magnetic moments in ferromagnetic CrTe plotted against the lattice constant a. *Inset*: Electronic pressure as a function of lattice constant (Reproduced from [41])

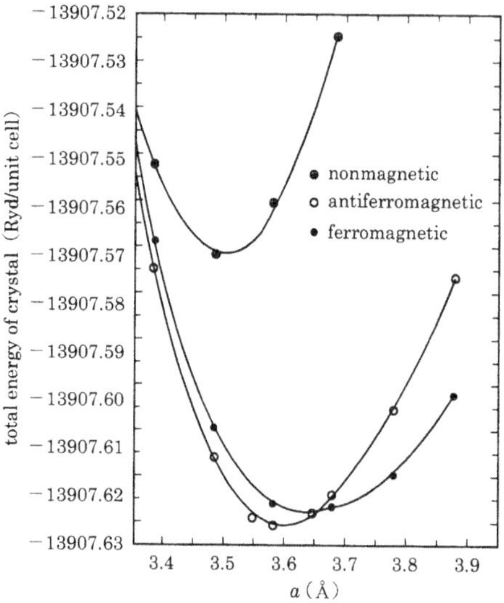

Fig. 4.32. Total energy for ferromagnetic, antiferromagnetic, and nonmagnetic CrSe (Reproduced from [41])

$\times 10^{-3}$ kbar^{-1}. The calculated compressibility evaluated by the FLAPW calculation is slightly smaller than the value 2.6×10^{-3} kbar^{-1} evaluated by the APW calculation. Pressure coefficient of magnetic moment is calculated as $\partial M/\partial P = -1.33\times 10^{-2}\mu_B$ kbar^{-1} at $a = 3.981$ Å, and then the magnetic moment decreases at the rate $k = -\partial \ln M/\partial P = -4.21 \times 10^{-3}$ kbar^{-1}, which is small but in the similar order in comparison with the observed value -12×10^{-3} kbar^{-1} [42].

We also carried out the total energy calculations for CrSe and CrS. As an example, we show the relation between E_{Total} and the a-axis of CrSe in Fig. 4.32. The total energy calculations show that the antiferromagnetic state is most stable even under high pressure both for CrSe and CrS.

4.6 Magnetic Ordering and Instability of Paramagnetic State

4.6.1 Double-Helical Magnetic Ordering of MnP-Type Compounds

The total energy calculation, as described in Sect. 4.5, makes it possible to clarify which magnetic state occurs in a material with itinerant electrons. It is, however, a hard task to find out the most stable magnetic state from the calculations of total energies, especially in a material with a lot of magnetic atoms in a unit cell because we need to carry out total energy calculations for many possible magnetic orderings. In addition, some of transition metal compounds such as CrAs explained in Part I have helical magnetic ordering. If wavevector of the helical ordering and lattice vectors are incommensurate, it is impossible to carry out band calculation.

An alternative way to search for a possible magnetic state is to study instability of paramagnetic state. In this section, we describe how we can investigate the instability of paramagnetic state on the basis of paramagnetic band calculations as well as the Hubbard model [44].

The transition metal pnictides such as FeAs and CrAs with the MnP-type crystal structure show the double-helical magnetic ordering. Magnetic transition temperature is 77 K for FeAs and 250 K for CrAs. We describe later the procedure to investigate instability of the paramagnetic state for the case of FeAs.

As shown in Fig. 4.33, the unit cell of FeAs (orthorhombic MnP-type structure) contains four Fe atoms and four As atoms. Paying attention only to Fe atoms and neglecting the difference in As atoms surrounding the respective Fe atoms, we can take a reduced unit cell which contains only two Fe atoms. The thick lines in Fig. 4.33 shows the reduced unit cell containing two Fe atoms labeled as "a" and "b".

Experimental studies have revealed that the helical ordering of FeAs is described by a wavevector $\boldsymbol{Q} = (0, 0, 0.375 \times 2\pi/c)$ (See also Tables 2.4 and 2.6). We note that Q of $0.375 \times 2\pi/c$ corresponds to wavelength $2.67b$. We also note that the lattice vector $\boldsymbol{c}$ here corresponds to $\boldsymbol{b}$ given in Table 2.6. It also has been known that the phase difference between magnetic moments at the a-site and the b-site in the same unit cell is $\phi = 140°$,[3] and magnitude of the magnetic moment of the Fe atom is $0.5\mu_B$ [31].

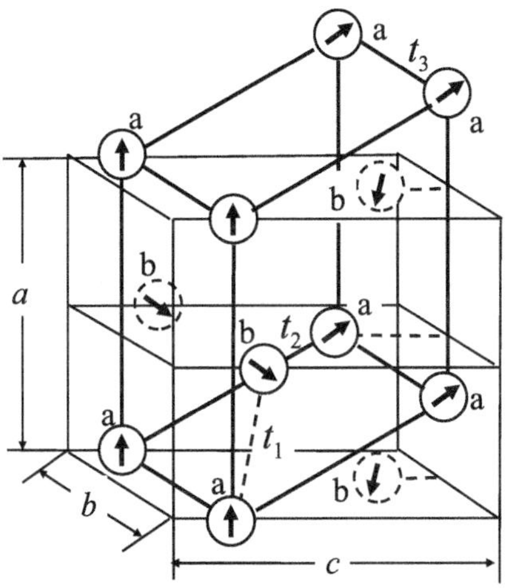

Fig. 4.33. The magnetic ordering of FeAs with the MnP-type structure. The thick lines denote the reduced unit cell (see the text). t_1 and t_2 denote transfer integrals between Fe atoms at different sublattices. t_3 is a transfer integral between Fe atoms at the same sublattices. See also Fig. 2.1 and Table 2.6

[3] ϕ is denoted by θ_{12} in Part I.

4.6.2 Instability of Paramagnetic State

We start from a Hubbard Hamiltonian with transfer integrals and intra-atomic Coulomb interaction. We assume that each Fe atom has a single orbital. For the transfer integral, we consider three types of interaction: t_3 between the same sublattice (a–a or b–b) and t_1 and t_2 between the different sublattices (a–b) as shown in Fig. 4.33.

By Fourier transformation and the Hartree–Fock approximation for the Coulomb interaction, we have the following expression of the Hubbard Hamiltonian:

$$
\begin{aligned}
\mathscr{H} = &\sum_{k\sigma} \left[T_1(\boldsymbol{k}) a^{\dagger}_{k\sigma} b_{k\sigma} + T_1^{*}(\boldsymbol{k}) b^{\dagger}_{k\sigma} a_{k\sigma} + T_3(\boldsymbol{k}) \left(a^{\dagger}_{k\sigma} a_{k\sigma} + b^{\dagger}_{k\sigma} b_{k\sigma} \right) \right] \\
&+ U \sum_{kq} \sum_{\sigma} \left[A_{q-\sigma} a^{\dagger}_{k+q\sigma} a_{k\sigma} + B_{q-\sigma} b^{\dagger}_{k+q\sigma} b_{k\sigma} \right] - NU \sum_{q} \left[A_{q+} A_{q-} + B_{q+} B_{q-} \right] \\
&- U \sum_{kq} \left[A^{+}_{q} a^{\dagger}_{k+q-} a_{k+} + B^{+}_{q} b^{\dagger}_{k+q-} b_{k+} + A^{-}_{-q} a^{\dagger}_{k+} a_{k+q-} + B^{+}_{-q} b^{\dagger}_{k+} b_{k+q-} \right] \\
&+ NU \sum_{q} \left[|A^{+}_{q}|^2 + |B^{+}_{q}|^2 \right],
\end{aligned}
\tag{4.22}
$$

where

$$
\begin{aligned}
A_{q\sigma} &= \frac{1}{N} \left\langle \sum_{k} a^{\dagger}_{k\sigma} a_{k+q\sigma} \right\rangle, \quad B_{q\sigma} = \frac{1}{N} \left\langle \sum_{k} b^{\dagger}_{k\sigma} b_{k+q\sigma} \right\rangle, \\
A^{\pm}_{\pm q} &= \frac{1}{N} \left\langle \sum_{k} a^{\dagger}_{k\pm} a_{k\pm q\mp} \right\rangle, \quad B^{\pm}_{\pm q} = \frac{1}{N} \left\langle \sum_{k} b^{\dagger}_{k\pm} b_{k\pm q\mp} \right\rangle,
\end{aligned}
\tag{4.23}
$$

and

$$
\begin{aligned}
T_1(\boldsymbol{k}) &= t_1 \sum_{j} \mathrm{e}^{\mathrm{i}\boldsymbol{k}\cdot(\boldsymbol{r}_i - \boldsymbol{r}_j - \boldsymbol{\tau})} + t_2 \sum_{j} \mathrm{e}^{\mathrm{i}\boldsymbol{k}\cdot(\boldsymbol{r}_i - \boldsymbol{r}_j - \boldsymbol{\tau})} \\
T_3(\boldsymbol{k}) &= t_3 \sum_{j} \mathrm{e}^{\mathrm{i}\boldsymbol{k}\cdot(\boldsymbol{r}_i - \boldsymbol{r}_j)}.
\end{aligned}
\tag{4.24}
$$

In (4.24), $\boldsymbol{r}_i$ denotes position of an a-atom in the ith unit cell, and $\boldsymbol{r}_j + \boldsymbol{\tau}$ position of a b-atom in the jth unit cell.

Using this Hamiltonian, we investigate instability of paramagnetic state at 0 K against an external magnetic field modulated by a wavevector $\boldsymbol{q}$. In order to do it, first we add Zeeman terms to the Hamiltonian. Next, by solving equations of motion of the quantities $A_{q\sigma}$, etc., we have the following relation between $\boldsymbol{q}$-dependent magnetic fields and magnetic moments:

$$\begin{pmatrix} H^a(\boldsymbol{q}) \\ H^b(\boldsymbol{q}) \end{pmatrix} = \begin{pmatrix} \bar{\chi}^{aa}(\boldsymbol{q}) & \bar{\chi}^{ab}(\boldsymbol{q}) \\ \bar{\chi}^{ba}(\boldsymbol{q}) & \bar{\chi}^{bb}(\boldsymbol{q}) \end{pmatrix} \begin{pmatrix} M^a(\boldsymbol{q}) \\ M^b(\boldsymbol{q}) \end{pmatrix} \equiv \bar{\chi}(\boldsymbol{q}) \begin{pmatrix} M^a(\boldsymbol{q}) \\ M^b(\boldsymbol{q}) \end{pmatrix}. \tag{4.25}$$

In this equation, $M^a(\boldsymbol{q}) = \mu_B(A_{\boldsymbol{q}-} - A_{\boldsymbol{q}+})$ and $M^b(\boldsymbol{q}) = \mu_B(B_{\boldsymbol{q}-} - B_{\boldsymbol{q}+})$ are Fourier components of spin density at the a- and b-sites, respectively. $H^a(\boldsymbol{q})$ and $H^b(\boldsymbol{q})$ denote $\boldsymbol{q}$-dependent magnetic fields operating on $M^a(\boldsymbol{q})$ and $M^b(\boldsymbol{q})$, respectively. $\bar{\chi}^{aa}(\boldsymbol{q})$, etc., are elements of the inverse magnetic susceptibility tensor. These quantities are given by

$$\bar{\chi}^{aa}(\boldsymbol{q}) = \bar{\chi}^{bb}(\boldsymbol{q}) = -\frac{1}{2\mu_B^2} \frac{\Gamma_1(\boldsymbol{q}) + U\left[\Gamma_1(\boldsymbol{q})^2 - |\Gamma_2(\boldsymbol{q})|^2\right]}{\Gamma_1(\boldsymbol{q})^2 - |\Gamma_2(\boldsymbol{q})|^2}, \tag{4.26a}$$

$$\bar{\chi}^{ab}(\boldsymbol{q}) = \bar{\chi}^{ba*}(\boldsymbol{q}) = \frac{1}{2\mu_B^2} \frac{\Gamma_2(\boldsymbol{q})^*}{\Gamma_1(\boldsymbol{q})^2 - |\Gamma_2(\boldsymbol{q})|^2}, \tag{4.26b}$$

where

$$\Gamma_1(\boldsymbol{q}) = \frac{1}{4N} \sum_{\boldsymbol{k}} \left[\chi_0^{\alpha\alpha}(\boldsymbol{k},\boldsymbol{q}) + \chi_0^{\beta\beta}(\boldsymbol{k},\boldsymbol{q}) + \chi_0^{\alpha\beta}(\boldsymbol{k},\boldsymbol{q}) + \chi_0^{\beta\alpha}(\boldsymbol{k},\boldsymbol{q})\right] \tag{4.27a}$$

and

$$\Gamma_2(\boldsymbol{q}) = \frac{1}{4N} \sum_{\boldsymbol{k}} \frac{T_1^*(\boldsymbol{k}+\boldsymbol{q}) T_1(\boldsymbol{k})}{|T_1(\boldsymbol{k}+\boldsymbol{q})|\,|T_1(\boldsymbol{k})|} \times \left[\chi_0^{\alpha\alpha}(\boldsymbol{k},\boldsymbol{q}) + \chi_0^{\beta\beta}(\boldsymbol{k},\boldsymbol{q}) - \chi_0^{\alpha\beta}(\boldsymbol{k},\boldsymbol{q}) - \chi_0^{\beta\alpha}(\boldsymbol{k},\boldsymbol{q})\right]. \tag{4.27b}$$

In these expressions, α and β denote bands, and $\chi_0^{\alpha\alpha}(\boldsymbol{k},\boldsymbol{q})$, etc., are bare electronic susceptibilities which are given by

$$\chi_0^{\mu\nu} = \frac{f(E_{\boldsymbol{k}\mu}) - f(E_{\boldsymbol{k}+\boldsymbol{q}\nu})}{E_{\boldsymbol{k}\mu} - E_{\boldsymbol{k}+\boldsymbol{q}\nu}},$$

where $E_{\boldsymbol{k}}^{\mu}$ denotes the band energy obtained by diagonalizing the Hamiltonian. When paramagnetic state is unstable against an infinitesimal $\boldsymbol{q}$-dependent magnetic field, a relation

$$\det[\bar{\chi}(\boldsymbol{q})] = \bar{\chi}^{aa}(\boldsymbol{q})^2 - |\bar{\chi}^{ab}(\boldsymbol{q})|^2 = 0 \tag{4.28}$$

must be satisfied. Substituting (4.26a)–(4.27b) into (4.28), we can evaluate a critical value of the Coulomb interaction U_c as a function of the wavevector $\boldsymbol{q}$. If $U \geq U_c$, paramagnetic state is unstable, resulting in occurrence of a magnetic order. When U_c takes the minimum value for a wavevector $\boldsymbol{q} = \boldsymbol{Q}$, we may say that a magnetic order described by the wavevector $\boldsymbol{Q}$ is most likely to realize.

Substituting $H^a(\boldsymbol{q}) = H^b(\boldsymbol{q}) = 0$ into (4.25), we have a relation $\bar{\chi}^{aa}(\boldsymbol{q})M^a(\boldsymbol{q}) + \bar{\chi}^{ab}(\boldsymbol{q})M^b(\boldsymbol{q}) = 0$. When $\boldsymbol{q} = \boldsymbol{Q}$, we see that $|\bar{\chi}^{aa}(\boldsymbol{q})|$ is equal to $|\bar{\chi}^{ab}(\boldsymbol{q})|$. Therefore, absolute value of $M^a(\boldsymbol{Q})$ is equal to the absolute value of $M^b(\boldsymbol{Q})$. In the real space, magnetic moments at the a- and the b-sites are written as $M^a(\boldsymbol{r}) = M^a(\boldsymbol{Q})\mathrm{e}^{-\mathrm{i}\boldsymbol{Q}\cdot\boldsymbol{r}}$ and $M^b(\boldsymbol{r}) = M^b(\boldsymbol{Q})\mathrm{e}^{-\mathrm{i}\boldsymbol{Q}\cdot\boldsymbol{r}}$, respectively.

From these expressions, the phase difference ϕ between the moments of the a-site and the b-site is defined by the following equation:

$$\frac{M^b(\boldsymbol{r}_i+\boldsymbol{\tau})}{M^a(\boldsymbol{r}_i)} = \frac{M^b(\boldsymbol{Q})}{M^a(\boldsymbol{Q})}\mathrm{e}^{-\mathrm{i}\boldsymbol{Q}\cdot\boldsymbol{\tau}} \equiv \mathrm{e}^{\mathrm{i}\phi}. \tag{4.29}$$

Since $\bar{\chi}^{aa}(\boldsymbol{Q})$ is a real number and the phase factor of $\bar{\chi}^{ab}(\boldsymbol{Q})$ is derived from $\Gamma_2(\boldsymbol{Q})^*$, ϕ the phase difference between moments at a-site and adjacent b-site is given by

$$\phi = \pi + \arg\left[\Gamma_2(\boldsymbol{Q})\right] + \arg\left[\mathrm{e}^{-\mathrm{i}\boldsymbol{Q}\cdot\boldsymbol{\tau}}\right]. \tag{4.30}$$

We carried out calculations using parameters $t_1 = -0.07\,\mathrm{Ry}$, $t_2/|t_1| = -1.0$, $t_3/|t_1| = -0.2$, and electron number $n = 1.1$. These values were chosen so that they reproduce as nicely as possible the features of Fermi surfaces and density of states of FeAs which were obtained by the APW band calculation.

The calculated results are shown in Figs. 4.34 and 4.35. As shown in Fig. 4.34, U_c takes the minimum value for $Q = 0.4 \times 2\pi/c$, which means that the paramagnetic state is most unstable against magnetic ordering modulated with this wavevector. We see that this value of $\boldsymbol{Q}$ agrees well with the experimental value $Q = 0.375 \times 2\pi/c$. Furthermore, as shown in Fig. 4.35, the phase difference ϕ takes 158° at $Q = 0.4 \times 2\pi/c$. This value of ϕ is also close to the experimental value $\phi = 140°$.

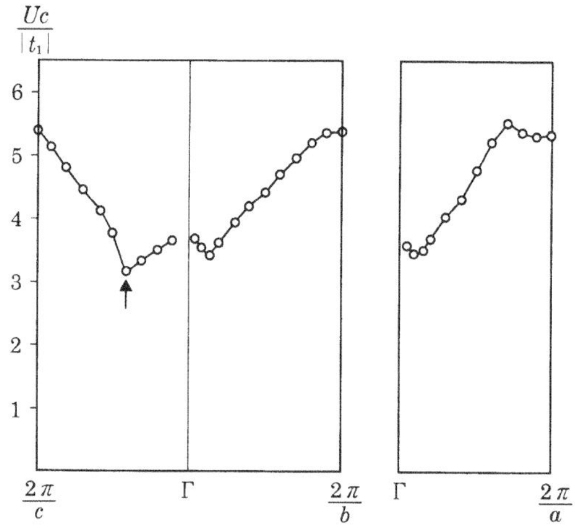

Fig. 4.34. Critical value of Coulomb interaction U_c evaluated as a function of wavevector $\boldsymbol{q}$. As shown by the *arrow*, U_c takes the minimum for $Q = 0.4 \times 2\pi/c$

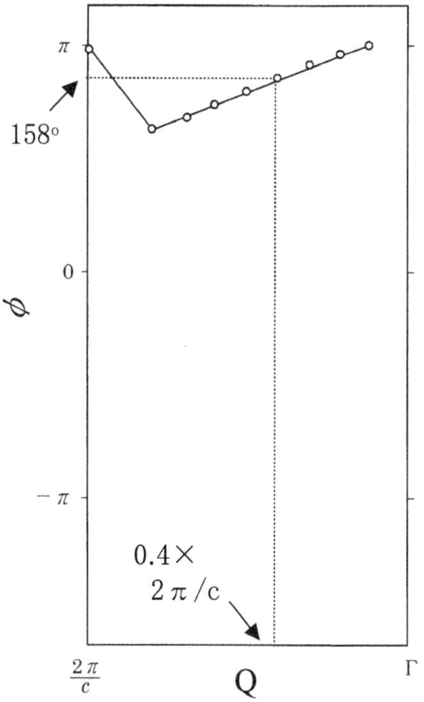

Fig. 4.35. Phase difference ϕ between spins at the a-site and b-site calculated as a function of $\boldsymbol{q}$. For $Q = 0.4 \times 2\pi/c$, $\phi = 158°$. This value of ϕ is close to the observed value $\phi = 140°$

4.6.3 Energy of Double-Helical Spin Density Wave State

Next, we investigate the double-helical spin density wave (DHSDW) state and evaluate electronic energy with the Hubbard Hamiltonian used in the instability study. In order to describe a DHSDW state modulated by a wavevector $\boldsymbol{Q}$, we neglect $\boldsymbol{q}$-terms except the one $\boldsymbol{q} = \boldsymbol{Q}$ in (4.22), and have

$$\mathscr{H} = \sum_{\boldsymbol{k}} \Psi^{\dagger}(\boldsymbol{k},\boldsymbol{Q})\hat{\mathscr{H}}(\boldsymbol{k},\boldsymbol{Q})\Psi(\boldsymbol{k},\boldsymbol{Q}) + \frac{NUn^2}{2} + NU\left(|A_{\boldsymbol{Q}}^{+}|^2 + |B_{\boldsymbol{Q}}^{+}|^2\right). \tag{4.31}$$

The $\hat{\mathscr{H}}(\boldsymbol{k},\boldsymbol{Q})$ and $\Psi(\boldsymbol{k},\boldsymbol{Q})$ in this equation are expressed by the following equations:

$$\hat{\mathscr{H}}(\boldsymbol{k},\boldsymbol{Q}) = \begin{pmatrix} T_3(\boldsymbol{k}) & T_1(\boldsymbol{k}) & -UA_{-\boldsymbol{Q}}^{-} & 0 \\ T_1^*(\boldsymbol{k}) & T_3(\boldsymbol{k}) & 0 & -UB_{-\boldsymbol{Q}}^{-} \\ -UA_{\boldsymbol{Q}}^{+} & 0 & T_3(\boldsymbol{k}+\boldsymbol{Q}) & T_1(\boldsymbol{k}+\boldsymbol{Q}) \\ 0 & -UB_{\boldsymbol{Q}}^{+} & T_1^*(\boldsymbol{k}+\boldsymbol{Q}) & T_3(\boldsymbol{k}+\boldsymbol{Q}) \end{pmatrix} \tag{4.32}$$

and

$$\Psi^{\dagger}(\boldsymbol{k},\boldsymbol{Q}) = \left[a^{\dagger}_{\boldsymbol{k}+}, b^{\dagger}_{\boldsymbol{k}+}, a^{\dagger}_{\boldsymbol{k}+\boldsymbol{Q}-}, b^{\dagger}_{\boldsymbol{k}+\boldsymbol{Q}-}\right]. \tag{4.33}$$

Total energy of the DHSDW state is given by

$$E_{\mathrm{SDW}} = \sum_{\boldsymbol{k}} \sum_{\mu=1}^{4} E_{\boldsymbol{k}\mu} f(E_{\boldsymbol{k}\mu}) + \frac{NUn^2}{2} + NU\left(|A^{+}_{\boldsymbol{Q}}|^2 + |B^{+}_{\boldsymbol{Q}}|^2\right), \tag{4.34}$$

where $E_{\boldsymbol{k}\mu}$ is eigen energies evaluated by diagonalizing the Hamiltonian, $f(E_{\boldsymbol{k}\mu})$ the Fermi–Dirac distribution function, $A^{+}_{\boldsymbol{Q}}$ and $B^{+}_{\boldsymbol{Q}}$ Fourier component of spin density at the a- and b-atoms, respectively. The phase difference ϕ between spins at the a-site and the b-site is given by

$$\frac{B^{+}_{\boldsymbol{Q}}}{A^{+}_{\boldsymbol{Q}}} \mathrm{e}^{-\mathrm{i}\boldsymbol{Q}\cdot\boldsymbol{\tau}} = \mathrm{e}^{\mathrm{i}\phi}. \tag{4.35}$$

We obtained $A^{+}_{\boldsymbol{Q}}$ and $B^{+}_{\boldsymbol{Q}}$ by self-consistent calculations, and evaluated the total energy E_{SDW} at 0 K as a function of $\boldsymbol{Q}$ and U.

In Fig. 4.36, E_{SDW} is plotted as a function of wavevector Q for the two values of $U/|t_1|$. The total energy E_{SDW} takes the smallest value at $Q = 0.4 \times (2\pi/c)$. This result corresponds to the result of the instability study described earlier.

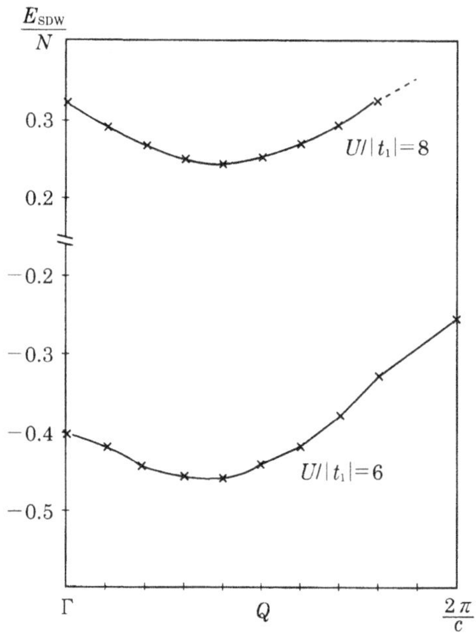

Fig. 4.36. Energy of the DHSDW state calculated as a function of wavevector $\boldsymbol{Q}$ for two values of $U/|t_1|$. Corresponding to the result of the instability study, the energy takes the minimum value at $Q = 0.4 \times 2\pi/c$

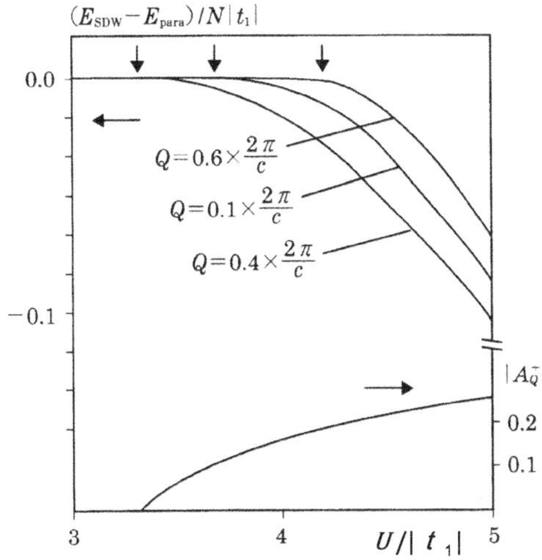

Fig. 4.37. Upper part: $E_{\text{SDW}} - E_{\text{para}}$ plotted as a function of $U/|t_1|$. For the curves of each value of Q, in the region of $U/|t_1|$ larger than the value denoted by the arrow, E_{SDW} is smaller than E_{para}. Lower part: $|A_Q^+|$ plotted as a function of $U/|t_1|$ calculated for $Q = 0.4 \times 2\pi/c$

We show energy difference between the DHSDW state and paramagnetic state, $E_{\text{SDW}} - E_{\text{para}}$, at 0 K in Fig. 4.37. The vertical arrows indicate the values of U, where E_{SDW} is equal to E_{para} for the respective values of Q shown in the figure. These values of U corresponds to U_{c} shown in (4.28). Since U_{c} takes the smallest value for $Q = 0.4 \times (2\pi/c)$ as shown in Fig. 4.37, the calculated energies of the DHSDW state is consistent with the result of instability study. The phase difference ϕ between spins in the same unit cell is evaluated from (4.35) as $\phi = 153$. The value of $|A_Q^+|$ for $Q = 0.4 \times (2\pi/c)$ is shown in lower part of Fig. 4.37 as a function of U. Since the magnetic moment per atom is given by $2|A_Q^+|$, we see that a parameter $U/|t_1| = 4.8$ reproduces the experimental value of magnetic moment $0.5\mu_{\text{B}}$/Fe, therefore we can estimate the magnitude of the Coulomb interaction to be $U = 4.6$ eV by using $t_1 = -0.07$ Ry.

From the discussion described in this section, we see that the double-helical magnetic ordering of FeAs is well explained by the itinerant picture based on the multi-band Hubbard model whose parameters (transfer integrals and electron number) are obtained so as to reproduce results of the APW band calculations. This theory is widely applicable not only to other MnP-type compounds which has the double-helical ordering but also to compounds with several magnetic atoms in a unit cell.

4.7 Phase Transition from the NiAs-Type to the MnP-Type Structure

The structural phase transition between the NiAs-type and the MnP-type structures, which is easily induced by temperature change or by composition change of compound, is one of the characteristic properties of the transition metal pnictides and chalcogenides with the NiAs-type structure. Correlation between magnetic properties and the crystal structure is an interesting problem which has not been fully understood yet. In transition metal chalcogenides, VS takes the MnP-type and the NiAs-type structures below and above $T_t = 850\,\mathrm{K}$, respectively. $Ti_{0.95}Se$ takes the NiAs-type structure, whereas $Ti_{1.05}Se$ takes the MnP-type structure at room temperature. $Ti_xV_{1-x}S$ takes the NiAs-type for $x > 0.66$ and the MnP-type for $x < 0.66$. As for the pnictides, MnAs, CrAs, and CoAs show the structural phase transition between the NiAs-type structure above T_t and the MnP-type crystal structure below T_t with $T_t = 398\,\mathrm{K}$, 1,100 K, 1,250 K, respectively. Only MnAs, with decreasing temperature from $T_t = 398\,\mathrm{K}$, shows the second transformation from the MnP-type to the NiAs-type structures at $T_t = 318\,\mathrm{K}$, where magnetic phase transition to the ferromagnetic order also occurs discontinuously (see also Fig. 3.17, etc.). In contrast to MnAs, etc., NiAs does not show the transformation to the MnP-type structure, however the de Haas–van Alfen effect and the Fermi surface suggest an existence of a structural transformation at very low temperatures as noted in Sect. 4.4.

Various measurements carried out for various mixed compounds revealed that the structural transformation between the NiAs-type and the MnP-type structures occurs in the following compounds: $Cr_{1-x}Mn_xAs$ (all x), $Cr_{1-x}Co_xAs$ (all x) , $Fe_{1-x}Co_xAs$ ($0.8 \le x \le 1.0$), $Mn_{1-x}Ni_xAs$ ($0 \le x \le 0.58$), $Mn_{1-x}Co_xAs$ ($0 \le x \le 0.2$, $0.5 \le x \le 1.0$), $MnAs_{1-x}P_x$ ($0 \le x \le 0.28$) See also Sect. 3.2. In contrast to these arsenides, all the antimonides investigated so far do not show the transformation to the MnP-type structure.

Taking account of the fact that the phase transition occurs in MnAs, CrAs, CoAs, and VS but not in MnSb, CrSb, NiAs, and TiS, we investigated mechanism of the structural phase transition by studying instability of the NiAs-type structure. This is a microscopic theory based on band structures obtained by the first principle calculations, that is, no fitting parameters are used.

4.7.1 Electron–Lattice Interaction Coefficient

Change of energy of an electronic system ΔF due to lattice deformation is described by a phonon normal mode $\boldsymbol{Q}_{q\lambda}$ ($\boldsymbol{q}$ is a wavevector, λ denotes mode) is given by

$$\Delta F = -\frac{1}{2}\chi(\boldsymbol{q}\lambda)\left|\boldsymbol{Q}_{q\lambda}\right|^2. \tag{4.36}$$

$\chi(\boldsymbol{q}\lambda)$ in this equation is a generalized susceptibility, which is given by

$$\chi(\boldsymbol{q}\lambda) = -\frac{1}{N}\sum_{\nu\nu'}\sum_{\alpha\beta}\frac{1}{\sqrt{M_\nu M_{\nu'}}}\varepsilon^\alpha(\boldsymbol{q}\lambda,\nu)\,\varepsilon^{\beta^*}(\boldsymbol{q}\lambda,\nu')\,\chi^{\alpha\beta}(\nu\nu',\boldsymbol{q}) \tag{4.37}$$

with

$$\chi^{\alpha\beta}(\nu\nu',\boldsymbol{q}) = 2\sum_{nn'}\sum_{\boldsymbol{k}} I^{\nu\alpha}_{n\boldsymbol{k},n'\boldsymbol{k}+\boldsymbol{q}}{}^{*} I^{\nu'\beta}_{n\boldsymbol{k},n'\boldsymbol{k}+\boldsymbol{q}} \frac{f(E^0_{n\boldsymbol{k}}) - f(E^0_{n'\boldsymbol{k}+\boldsymbol{q}})}{E^0_{n\boldsymbol{k}} - E^0_{n'\boldsymbol{k}+\boldsymbol{q}}}. \tag{4.38}$$

In these equations, M_ν is mass of the ν atom, $\varepsilon(q\lambda,\nu)$ is the polarization vector of phonon, E^0_{nk} is the band energy in the absence of lattice distortion, and $f(E^0_{nk})$ is the Fermi–Dirac distribution function.

Displacement of a ν atom in the α-direction brings about change of crystalline potential ΔV, resulting in coupling between eigen states $|n\boldsymbol{k}\rangle$ and $|n'\boldsymbol{k}+\boldsymbol{q}\rangle$ of the system. The coefficient $I^{\nu\alpha}_{n\boldsymbol{k},n'\boldsymbol{k}+\boldsymbol{q}}$ in (4.38) denotes magnitude of the coupling between states $\Psi_{n\boldsymbol{k}}$ and $\Psi_{n'\boldsymbol{k}+\boldsymbol{q}}$ due to the potential change ΔV, and is called an electron–lattice interaction coefficient, which is given by

$$I^{\nu\alpha}_{n\boldsymbol{k},n'\boldsymbol{k}+\boldsymbol{q}} = \left\langle \Psi_{n\boldsymbol{k}} \middle| \Delta V \middle| \Psi_{n'\boldsymbol{k}+\boldsymbol{q}} \right\rangle . \tag{4.39}$$

An expression of the electron–lattice interaction coefficient based on the APW formalism is given in [45].

4.7.2 Tendency of Structural Transformation from the NiAs-Type to the MnP-Type

The structural transformation to the MnP-type is caused by freezing of a phonon mode M_4^- at the M point of the first Brillouin zone for the NiAs-type structure. The normal coordinate of the M_4^- mode is expressed as

$$c_1(z_1+z_2)+c_2(x_1-x_2)+c_3(z_3-z_4), \tag{4.40}$$

where the subscripts 1 and 2 denote metal atoms, and 3 and 4 denote anion atoms. The symbol z is the displacement along the c-axis, and x (y) the displacement in the c-plane (see Fig. 4.38). As indicated in (4.40), metal atoms shift in the c-plane as well as along the c-axis, while anions only displace along the c-axis (see also Chap. 1).

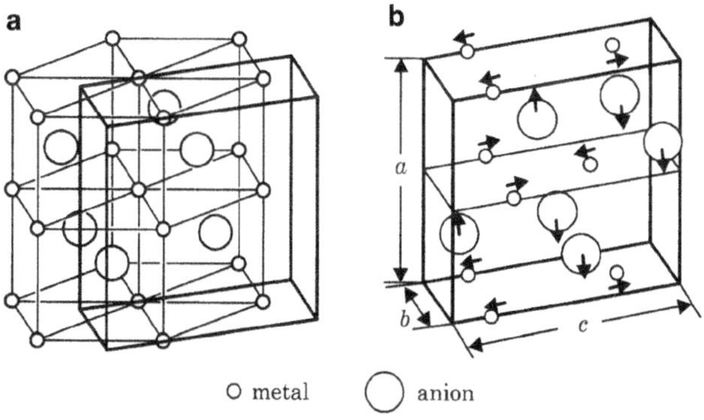

Fig. 4.38. (**a**) The NiAs-type structure. The *thick lines* show the unit cell corresponding to the MnP-type structure. (**b**) Unit cell of the MnP-type structure. The arrows indicate displacements of atoms which vanish for the NiAs-type structure. We note that magnitude of the displacement is exaggerated. The lattice parameters a, b, and c in this figure correspond to A, B, C in Fig. 1.4, respectively

We calculated matrix elements of the electron–lattice interaction $I^{\nu\alpha}_{nk,n'k+q}$ for $\boldsymbol{q} = \Gamma$M as a function of $\boldsymbol{k}$. We note that when a wavevector $\boldsymbol{k}$ is on the Σ-line (ΓM), $\boldsymbol{k}+\boldsymbol{q}$ is also on the Σ-line. There are four irreducible representations Σ_1, Σ_2, Σ_3, and Σ_4 in the group of symmetry on the Σ-line. Among product representations, only $\Sigma_1 \times \Sigma_4$ and $\Sigma_2 \times \Sigma_3$ satisfy compatibility relation with the M_4^- representation.

Therefore, matrix elements I's vanish except for the combination of states Σ_1- and Σ_4-bands or Σ_2- and Σ_3-bands. Furthermore, it is clear from (4.38) that the electron–lattice interaction reduces the free energy only when one of the states $|n\boldsymbol{k}\rangle$ and $|n'\boldsymbol{k}'\rangle$ is below the Fermi level and the other is above the Fermi level, and the reduction of free energy increases as the energy difference $E^0_{nk} - E^0_{n'k'}$ decreases.

MnAs, MnSb, CrAs, and CrSb

Dispersion curves along the Σ-line near the Fermi level calculated by using the APW method for MnAs, MnSb, CrAs, and CrSb are shown in Fig. 4.39. We see that the Σ_1-band and Σ_4-band lie near the Fermi level, so we calculated $I^{\nu\alpha}_{nk,n'k+q}$ for the states $\boldsymbol{k}$ of the Σ_1-band and $\boldsymbol{k}+\boldsymbol{q}$ of the Σ_4-band. Calculated electron–lattice interaction coefficients are plotted as a function of $\boldsymbol{k}$ along the Σ-line in Fig. 4.40. The symbols •, ×, and ○ denote the results for the displacements in the x-direction of a metal atom, in the z-direction of a metal atom, and in z-direction of an anion, respectively. In the hatched regions, on the $\boldsymbol{k}$-axis, the electron–lattice interaction contributes to the reduction of energy of electron system. Figure 4.41 shows the reduction of free energy $|I^{\nu\alpha}_{\Sigma_1 k,\Sigma_4 k+q}|^2/|E_{\Sigma_1 k} - E_{\Sigma_4 k+q}|$. When this quantity is large in the hatched region on the $\boldsymbol{k}$-axis shown in Fig. 4.41, large reduction of energy of the electron system due to the lattice distortions arises, therefore we can say that phase transition to the MnP-type structure is likely to occur. The results shown in Fig. 4.41 suggests that the phase transition to the MnP-type structure is expected to occur in MnAs and CrAs, but not in MnSb and CrSb.

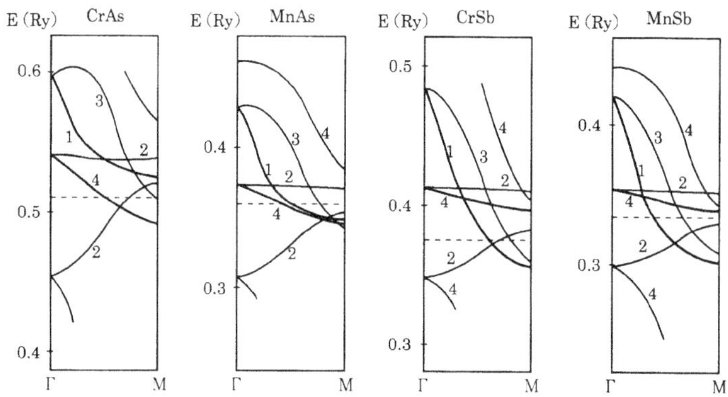

Fig. 4.39. Dispersion curves of the nonmagnetic bands of MnAs, MnSb, CrAs, CrSb along the Σ-line. The broken line denotes the Fermi level. The numbers on the curves 1, 2, 3, and 4 denote the irreducible representations $\Sigma_1, \Sigma_2, \Sigma_3$, and Σ_4 of the bands (Reproduced form [45] with permission)

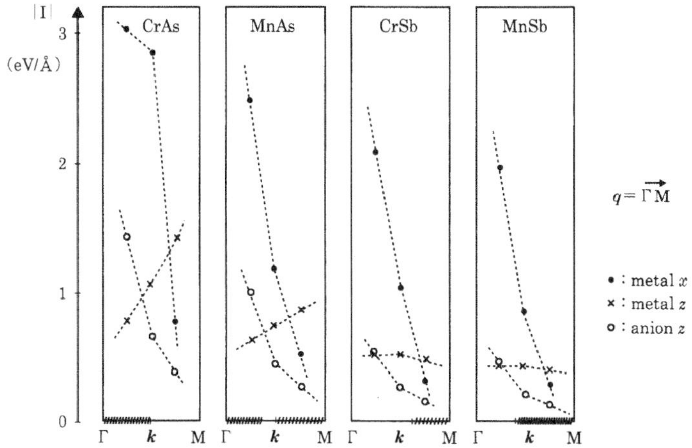

Fig. 4.40. Elecron–lattice matrix elements, $|I^{\nu\alpha}_{\Sigma_1 \boldsymbol{k},\Sigma_4 \boldsymbol{k}+\boldsymbol{q}}|$, for MnAs, MnSb, CrAs, CrSb as functions of the wavevector $\boldsymbol{k}$. The wavevector $\boldsymbol{q}$ is fixed to ΓM. The metal x-, metal z-, and anion z-components of the electron–lattice matrix elements represented by *filled circles*, *times*, and *open circles*, respectively. The *hatched regions* on the $\boldsymbol{k}$-axis denote, where the coupling of electronic states due to lattice distortion reduces electronic energy (Reproduced form [45] with permission)

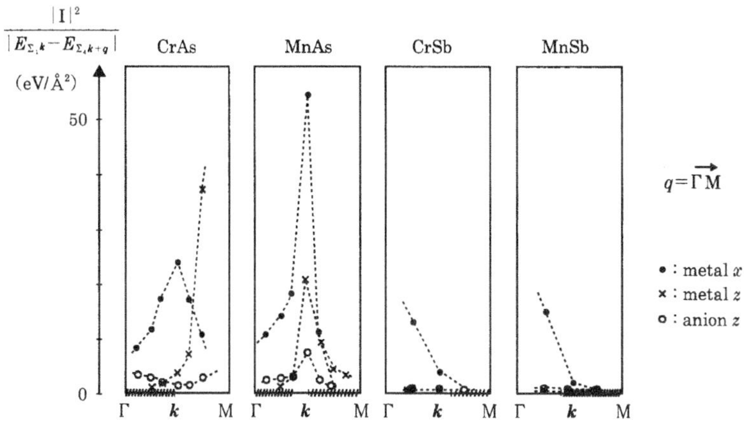

Fig. 4.41. Energy reduction $|I^{\nu\alpha}_{\Sigma_1 \boldsymbol{k},\Sigma_4 \boldsymbol{k}+\boldsymbol{q}}|^2/|E_{\Sigma_1 \boldsymbol{k}} - E_{\Sigma_4 \boldsymbol{k}+\boldsymbol{q}}|$ ($\boldsymbol{q} = \Gamma$M) calculated as functions of $\boldsymbol{k}$. The symbols *filled circles*, *times*, and *open circles* denote the metal x-, metal z-, and anion z-components of the MnP-type distortion (Reproduced with permission from [45])

We also see from the figure that the energy reduction due to the displacement of metal atoms in the x-direction is larger than that due to other two displacements. Observation for MnAs in the MnP-type structure has revealed that the displacement of metal atoms in the x-direction is the largest in the three displacements.

From the discussion mentioned earlier, we see that wavevector dependence of the electron–lattice interaction coefficients and dispersion curves near the Fermi level have a decisive role in the phase transition form the NiAs-type to the MnP-type structure.

CoAs and NiAs

Using the APW method, we calculated the band structure of CoAs and NiAs near the Fermi level for $\boldsymbol{k}$-points in the planes $k_z = 0$, $k_z = 1/3\Gamma$A, and $k_z = 2/3\Gamma$A [46]. As an example of calculations, we show dispersion curves along the lines parallel to the Σ-line (ΓM) on the plane of k_z =0 in Fig. 4.42. As we have noted earlier, the combination of Σ_1-band and Σ_4-band (or the Σ_2-band and Σ_3-band) contribute to the reduction of electronic energy due to the MnP-type lattice distortion. Only when one of the states is above the Fermi level and the other is below the Fermi level, the electron–lattice interaction reduces electronic energy. The hatched region in Fig. 4.42 indicates where the reduction of electronic energy occurs.

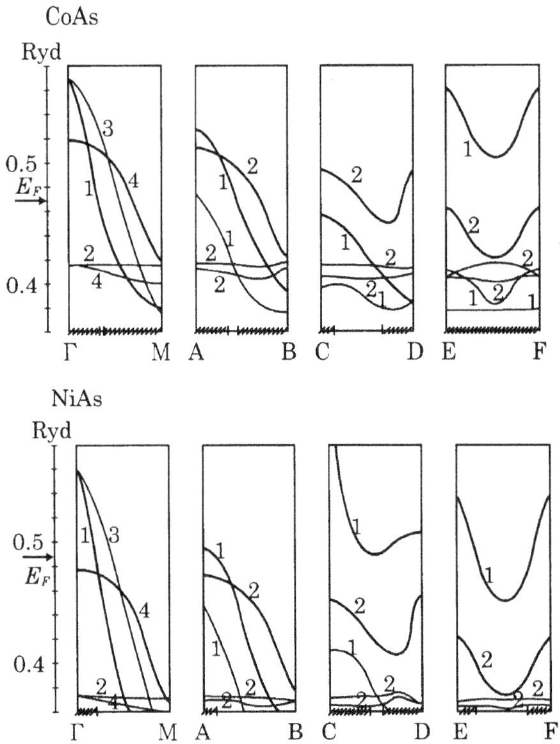

Fig. 4.42. Dispersion curves of the band of CoAs and NiAs near the Fermi level along the direction parallel to the ΓM. the *hatched regions* on the $\boldsymbol{k}$-axis mean where the energy reduction due to the electron–lattice interaction occurs (Reproduced with permission from [46])

By calculating $|I^{\nu\alpha}_{\Sigma_1 \boldsymbol{k},\Sigma_4 \boldsymbol{k}+\boldsymbol{q}}|^2/|E_{\Sigma_1 \boldsymbol{k}} - E_{\Sigma_4 \boldsymbol{k}+\boldsymbol{q}}|$, we found that reduction of electronic energy due to the MnP-type lattice distortion is large for CoAs but small for NiAs, which indicates that structural transformation to the MnP-type structure is likely to occur in CoAs but not in NiAs. This is consistent with observation. Calculated electron–lattice interaction coefficients as well as $|I^{\nu\alpha}_{\Sigma_1 \boldsymbol{k},\Sigma_4 \boldsymbol{k}+\boldsymbol{q}}|^2/|E_{\Sigma_1 \boldsymbol{k}} - E_{\Sigma_4 \boldsymbol{k}+\boldsymbol{q}}|$ are depicted in [46].

VS and TiSe

We carried out similar calculations for VS and TiSe. Dispersion curves near the Fermi level of these materials show that the electron–lattice interaction between the states $\boldsymbol{k}$ of the Σ_3-band and $\boldsymbol{k}+\boldsymbol{q}$ of the Σ_2-band is important. Calculated electron–lattice interaction coefficients show that reduction of free energy due to MnP-type lattice distortion is large for VS in comparison with that for TiSe, which suggests that VS has a possibility to deform to the MnP-type structure, however TiSe does not.

References

[1] K. Motizuki, in *Recent Advances in Magnetism of Transition Metal Compounds*, ed. by A. Kotani, N. Suzuki (World Scientific, Singapore, 1993), p. 26; K. Motizuki, J. Magn. Magn. Mater. **70**, 1 (1987)

[2] L.M. Sandratskii, R.F. Egorov, A.A. Berdyshev, Phys. Stat. Sol. (B) **103**, 511 (1981)

[3] Y. Yamaguchi, H. Watanabe, J. Magn. Magn. Mater. **31–34**, 619 (1983)

[4] R. Coehoorn, C. Haas, R.A. de Groot, Phys. Rev. B **31**, 1980 (1985)

[5] K. Nakada, Doctoral dissertation

[6] K.S. Liang, T. Chen, Solid State Commun. **23**, 975 (1977)

[7] M. Morifuji, K. Motizuki, J. Phys. Soc. Jpn. **57**, 3411 (1988); M. Morifuji, K. Motizuki, J. Magn. Magn. Mater. **70**, 70 (1987)

[8] Landolt-Börnstein III/27a, *Magnetic Properties of Pnictides and Chalcogenides*, ed. by K. Adachi, S. Ogawa (Springer, Berlin, 1989), p. 70

[9] A. Kallel, H. Boller, F. Bertaut, J. Phys. Chem. Solids **35**, 1139 (1974)

[10] A.I. Snow, Phys. Rev. **85**, 365 (1952)

[11] K. Selte, A. Kjekshus, W.A. Jamison, A.F. Andresen, J.E. Engebresen, Acta Chem. Scand. **25**, 1703 (1971)

[12] K. Selte, A. Kjekshus, A.F. Andresen, Acta Chem. Scand. **26**, 4188 (1972)

[13] T. Suzuki, H. Ido, J. Appl. Phys. **73**, 5686 (1993)

[14] S.Y. Savrasov, Full-Potential Program Package gLMTART 6.50 h, New Jersey Institute of Technology (2003)

[15] S.H. Vosko, L. Wilk, M. Nussair, Can. J. Phys. **58**, 1200 (1980)

[16] T. Kamimura; private communication

[17] K. Selte, H. Hjersing, A. Kjekshus, A.F. Andersen, P. Fischer, Acta Chem. Scand. A **29**, 695 (1975)

[18] V.A. Shanditsev, L.F. Vereschchagin, E.N. Yakovlev, N.P. Grazhdankina, T.I. Alaeva, Sov. Phys. Solid State **15**, 146 (1973)

[19] T. Kanomata, N. Suzuki, H. Nishihara, T. Kaneko, H. Kato, N. Fujii, M. Ishizuka, S. Endo, Physica B **284–288**, 1515 (2000); N. Suzuki, T. Kanomata, R. Konno, T. Kaneko, H. Yamauchi, K. Koyama, H. Nojiri, Y. Yamaguchi, M. Motokawa, J. Alloys Comp. **290**, 25 (1999)

[20] T. Eto, M. Ishizuka, S. Endo, T. Kanomata, T. Kikegawa, J. Alloys Comp. **315**, 16 (2001); M. Ishizuka, H. Kato, T. Kunise, S. Endo, T. Kanomata, H. Nishihara, J. Alloys Comp. **320**, 24 (2001)
[21] E. Kulatov, L. Vinokurova, K. Motizuki, in *Recent Advances in Magnetism of Transition Metal Compounds*, ed. by A. Kotani, N. Suzuki (World Scientific, Singapore, 1993), p. 56
[22] J.H. Schwartz, E.L. Hall, G.P. Felcher, J. Appl. Phys. **42**, 1621 (1971)
[23] T. Moriya, in *Spin Fluctuations in Itinerant Electron Magnetism*, Springer Series in Solid-State Science, vol. 56 (Springer, Berlin, 1985)
[24] T. Moriya, Y. Takahashi, J. Phys. Soc. Jpn. **45**, 397 (1978)
[25] K. Usami, T. Moriya, J. Magn. Magn. Mater. **20**, 171 (1980)
[26] K. Motizuki, K. Katoh, J. Phys. Soc. Jpn. **53**, 735 (1984)
[27] K. Katoh, K. Motizuki, J. Phys. Soc. Jpn. **53**, 3166 (1984)
[28] T. Suzuki, H. Ido, J. Phys. Soc. Jpn. **51**, 3149 (1982)
[29] V. Heine, Phys. Rev. **153**, 673 (1967)
[30] K. Motizuki, M. Morifuji, J. Phys. (Paris) **C8**, 189 (1988)
[31] K. Selte, A. Kjekshus, A.F. Andresen, Acta Chem. Scand. **26**, 3101 (1972); K. Selte, A. Kjekshus, A.F. Andresen, Acta Chem. Scand. **25**, 3277 (1971)
[32] Y. Takahashi, T. Moriya, J. Phys. Soc. Jpn. **46**, 1451 (1979)
[33] M. Morifuji, K. Motizuki, J. Magn. Magn. Mater. **70**, 70 (1987)
[34] K. Motizuki, J. Magn. Magn. Mater. **70**, 1 (1987)
[35] T. Nozue, H. Kobayashi, T. Kamimura, T. Kawakami, H. Harima, K. Motizuki, J. Phys. Soc. Jpn. **68**, 2067 (1999)
[36] H. Harima, A. Yanase, J. Phys. Soc. Jpn. **60**, 2718 (1991)
[37] K. Motizuki, T. Kawakami, M. Oohigashi, H. Harima, T. Nozue, H. Kobayashi, T. Kamimura, Physica B **284–288**, 1345 (2000)
[38] W.P. Ellis, R.C. Albers, J.W. Allen, Y. Laissailly, J.-S. Kang, Solid State Commun. **62**, 591 (1987)
[39] R. Vincent, R.L. Withers, Philos. Mag. Lett. **56**, 57 (1987)
[40] H. Kobayashi, M. Kageshima, N. Kimura, H. Aoki, M. Oohigashi, K. Motizuki, T. Kamimura, J. Magn. Magn. Mater. **272–276**, e247 (2004)
[41] T. Kawakami, N. Nakata, K. Motizuki, J. Magn. Magn. Mater. **196–197**, 629 (1999)
[42] M. Takagaki, T. Kawakami, N. Tanaka, M. Shirai, K. Motizuki, J. Phys. Soc. Jpn. **67**, 1014 (1998)
[43] H. Shoren, F. Ikemoto, K. Yoshida, N. Tanaka, K. Motizuki, Physica E **10**, 242 (2001)
[44] M. Morifuji, K. Motizuki, J. Magn. Magn. Mater. **90–91**, 740 (1990); M. Morifuji, Doctral Thesis
[45] K. Katoh, K. Motizuki, J. Phys. Soc. Jpn. **56**, 655 (1987)
[46] M. Morifuji, K. Motizuki, J. Phys. Soc. Jpn. **57**, 3411 (1988)

5

Itinerant Electron Theory of Magnetism of Cu_2Sb-Type Compounds

5.1 Crystal Structure and Magnetic Properties

The Cu_2Sb-type compounds that consist of Cr, Mn, Fe, and As or Sb take tetragonal crystal structure. Figure 5.1 shows the crystal structure and the first Brillouin zone of the Cu_2Sb-type structure. The unit cell contains four magnetic atoms denoted by (1, 2, 3, 4) and two As or Sb atom denoted by (5, 6). Magnetic atoms occupy the sites M(I) and M(II), which have different symmetry. The Cu_2Sb-type compounds have attracted much interest because they show various magnetic ordering such as ferromagnetic, ferrimagnetic, and antiferromagnetic orderings [1].

Magnetic structure with many magnetic atoms in a unit cell is described by two parameters $\boldsymbol{Q}$ and ϕ_{ij}. The vector $\boldsymbol{Q}$ (propagation vector) is a wavevector that specifies angle between total moment in a unit cell and that in an adjacent unit cell, and ϕ_{ij} is a relative angle between moments at the ith and the jth sites in a unit cell. The structural parameters and magnetic structure ($\boldsymbol{Q}$, ϕ_{ij}, and magnetic moments at the respective sites) are shown in Tables 5.1 and 5.2, respectively.

In CrMnAs and FeMnAs, which are mixed compounds between Mn_2As and Cr_2As or Fe_2As, a Cr or Fe atom occupies mainly the M(I)-site and a Mn atom occupies mainly the M(II)-site. In MnAlGe and MnGaGe, which also take the Cu_2Sb-type structure, Mn atoms occupy only the M(I)-site, and Al and Ga occupy the M(II)-site. The compounds MnAlGe and MnGaGe are ferromagnetic below $T_C = 503$ and 453 K, respectively. As the Mn atomic layers, as seen in Fig. 5.1, are separated by the layers of Al (or Ga) and Ge, these compounds are of interest as a quasi two-dimensional ferromagnet.

5.2 Band Structures of Cu_2Sb-Type Compounds and Magnetic and Optical Properties

Since early 1990s, Motizuki and her co-workers have carried out band calculations of the Cu_2Sb-type compounds for magnetic and nonmagnetic phases to investigate various magnetic properties as shown in Table 5.2 [2]. The self-consistent APW method

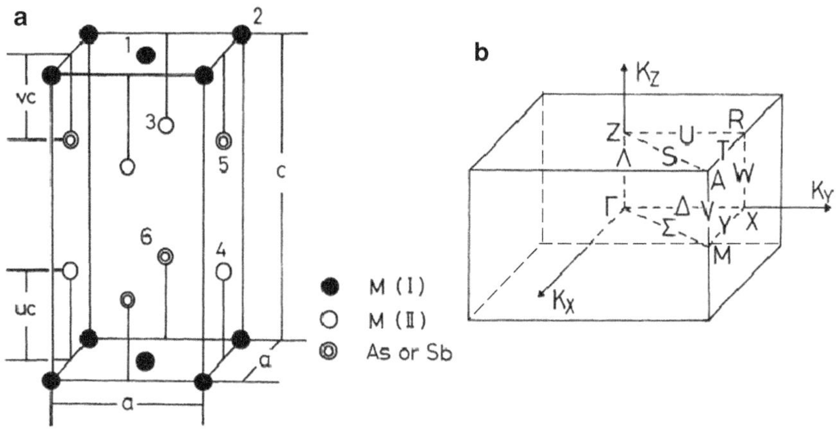

Fig. 5.1. (**a**) Crystal structure and (**b**) the first Brillouin zone of the Cu_2Sb-type structure

Table 5.1. Lattice parameters of Cu_2Sb-type compounds

	Lattice parameters			
	a (Å)	c (Å)	u_c	v_c
Cr_2As	3.60	6.34	0.325	0.275
CrMnAs	3.88	6.28	0.327	0.266
Mn_2As	3.78	6.25	0.33	0.265
MnFeAs	3.735	6.035	0.33	0.25
Fe_2As	3.627	5.973	0.33	0.265
Mn_2Sb	4.078	6.557	0.295	0.280
Cu_2Sb	3.99	6.09	0.27	0.30
MnAlGe	3.914	5.933	0.273	0.280
MnGaGe	3.963	5.895	0.29	0.29

Data are taken from [1]

and the LAPW method are employed [3] together with the muffin-tin approximation for crystal potential and the local density approximation for the exchange and correlation interaction. The formula of Gunnarsson and Lundqvist (GL) was also used for the LDA term [4]. As for relativistic effects, we neglected the spin–orbit interaction but scalar relativistic effects are taken into account. Densities of states calculated by the APW and LAPW methods for the nonmagnetic bands of the Cu_2Sb-type compounds were found to agree well with each other. APW band calculations including the spin–orbit interaction were carried out by Kulatov and his co-workers for the Cu_2Sb-type compounds. They also calculated optical conductivity from the band structure [5].

Table 5.2. Magnetic structure of Cu_2Sb-type compounds

		Magnetic structure				Magnetic moments (μ_B/atom)	
		Q	ϕ_{12}	ϕ_{23}	ϕ_{34}	M(I)	M(II)
Cr_2As	AF	$(0,0,\pi/c)$	π	π	0	0.40	1.34
CrMnAs	AF	(0,0,0)	π	π	π	0.41	3.14
Mn_2As	AF	$(0,0,\pi/c)$	0	π	π	2.2	4.1
MnFeAs	AF	$(0,0,\pi/c)$	0	0	π	0.2	3.6
Fe_2As	AF	$(0,0,\pi/c)$	0	0	π	1.28	2.05
Mn_2Sb	FI	(0,0,0)	0	π	0	2.13	3.87
Cu_2Sb	P	–	–	–	–	–	–
MnAlGe	F	(0,0,0)	0	0	0	1.70	–
MnGaGe	F	(0,0,0)	0	0	0	1.66	–

Data are taken from [1]

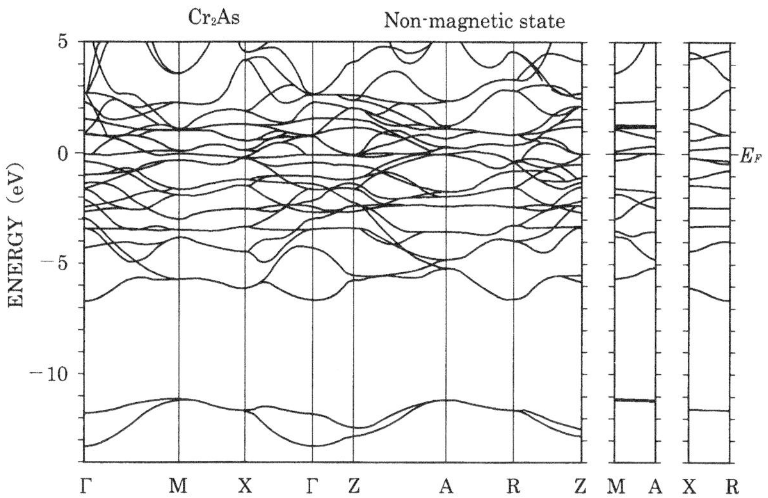

Fig. 5.2. Dispersion curves of nonmagnetic band of Cr_2As

5.2.1 Nonmagnetic State of Cr_2As, Mn_2As, Fe_2As, Mn_2Sb, CrMnAs, and FeMnAs

We calculated dispersion curves and densities of states for these compounds in non-magnetic state [6]. As an example, we show the dispersion curves of Cr_2As in Fig. 5.2. Two bands in the low energy region consist of As-4*s* orbitals. Above the gap, there are mixing bands between Cr-3*d* orbitals and As-4*p* orbitals. Total width of the *p*–*d* mixing bands is about 3–4 eV, which is large enough to regard the *d* electrons as itinerant electrons. Figure 5.3 shows the total and the partial densities of states of M_2As (M = Cr, Mn, and Fe) and Mn_2Sb. When we divide the *p*–*d* mixing band

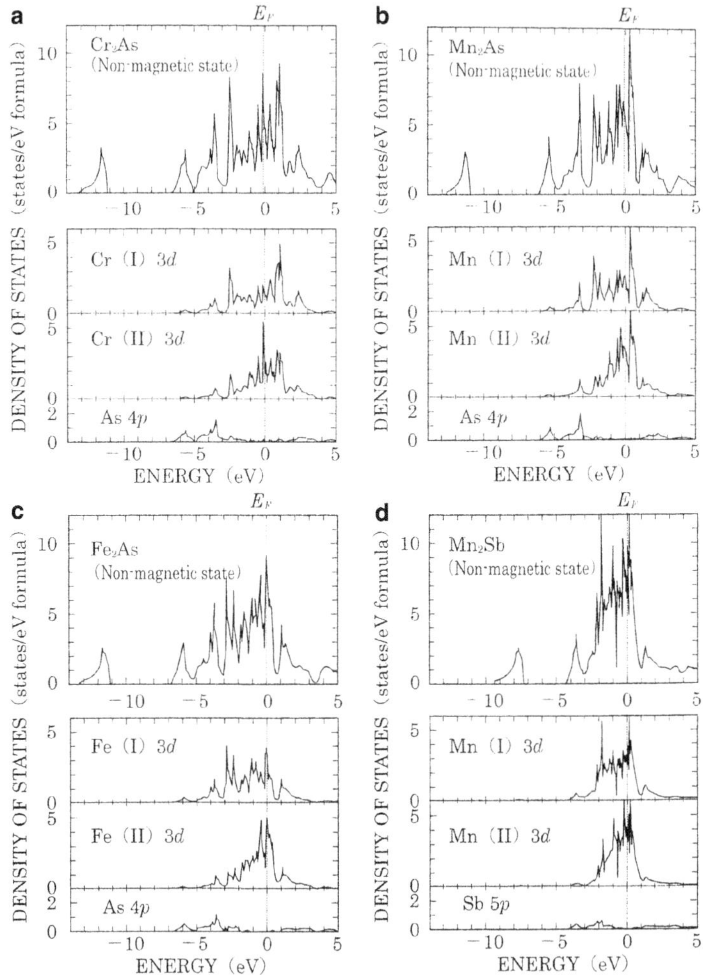

Fig. 5.3. Densities of states of nonmagnetic (**a**) Cr_2As, (**b**) Mn_2As, (**c**) Fe_2As and (**d**) Mn_2Sb

shown in Fig. 5.3 into three parts specified by (1), (2), and (3) from low energy side, it was revealed by our calculations that (1) and (3) have bonding and antibonding characters between M and As (or Sb), respectively, and (2) is regarded as nonbonding bands consisting almost of *d* obitals. In the nonbonding bands, *d*-orbitals at the M(I)-sites and M(II)-sites show strong hybridization as seen in Fig. 5.3.

It is generally said from Fig. 5.3 that the *d*-band arising from M(II)-sites is narrower than that from M(I)-sites, which indicates that *d* electrons at the M(II)-sites have weak itinerancy compared with *d* electrons at the M(I)-sites. Therefore, we expect that magnitude of magnetic moment at the M(II)-site is larger than that at the M(I)-site. Both calculations and experiments actually indicate that magnetic moment of M(II) is larger than that of M(I) as shown in Table 5.3.

Table 5.3. Magnetic moments of Cu_2Sb-type compounds evaluated by calculations

	Magnetic moments (μ_B)						
	Total		M(I)		M(II)		Anion
Cr_2As	–		0.34	(0.40)	1.37	(1.34)	0.03
Mn_2As	–		1.72	(2.2)	3.50	(4.1)	0.04
Fe_2As	–		1.01	(1.28)	2.11	(2.05)	0.03
Mn_2Sb	1.76	(1.74)	−2.11	(−2.13)	3.67	(3.87)	0.01
MnAlGe	1.81	(1.70)	1.90		−0.02		−0.06
MnGaGe	2.22	(1.66)	2.31		−0.02		−0.07

Observed moments are shown in the parenthesis

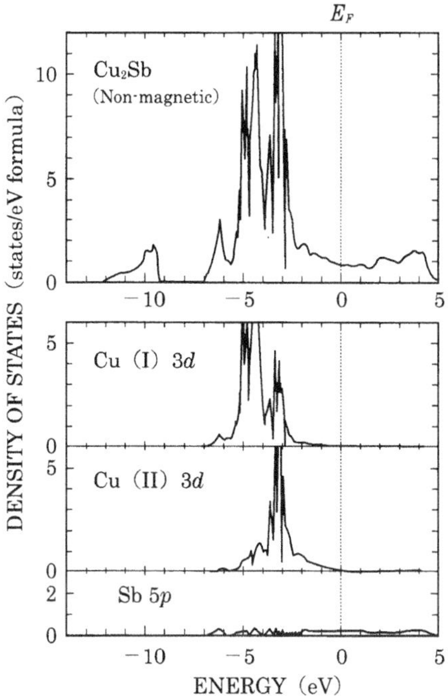

Fig. 5.4. Density of states of nonmagnetic Cu_2Sb

Fermi level lies in the nonbonding band. As seen in Fig. 5.3, Mn_2Sb has larger $\rho(E_F)$, the density of states at E_F, while compounds other than Mn_2Sb have smaller $\rho(E_F)$ than that of Mn_2Sb, which is consistent with the fact that Mn_2Sb is ferrimagnetic and others are antiferromagnetic.

Next, we show in Fig. 5.4 the density of states of Cu_2Sb which shows no magnetic ordering [7]. We see that hybridization between d-orbitals of M(I) and M(II) is weak, and density of states at Fermi level, $\rho(E_F)$, is very small.

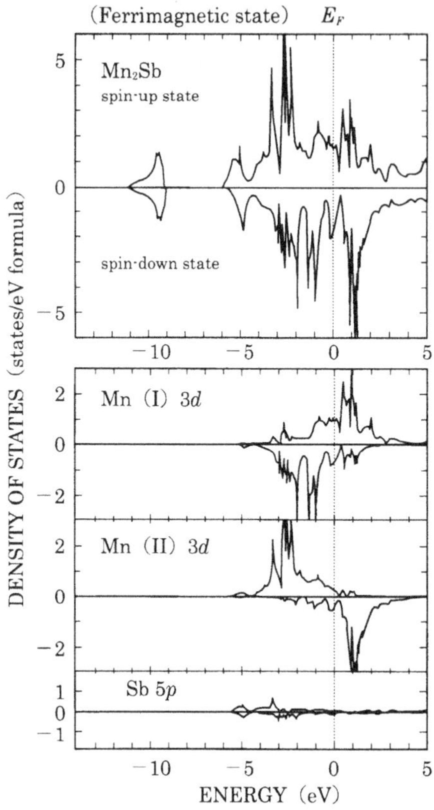

Fig. 5.5. Density of states of ferrimagnetic Mn_2Sb

5.2.2 Ferrimagnetic Band of Mn_2Sb

We show calculated density of states of ferrimagnetic Mn_2Sb in Fig. 5.5 [8]. Components of $3d$-orbitals of Mn(I) and Mn(II) and $5p$-orbitals of Sb inside muffin-tin spheres are also plotted. It is noted that the shape of the densities of states for the up-spin and down-spin bands are quite different from each other in the respective components, and also different from the densities of states in the nonmagnetic states. The band component of Mn(II) shows a large splitting between up- and down-spin bands, while the splitting of the band of Mn(I) is small, which results in that the magnetic moment of Mn(II) is about two times larger than that of Mn(I) (see Table 5.3). We also see that small magnetic moment is induced at the Sb-site in the direction of moment of Mn(II). Calculated total magnetic moment as well as those of Mn(I) and Mn(II) agrees well with experimental vales as seen in Table 5.3.

The ferrimagnetic band of Mn_2Sb is also calculated by Haas and his co-workers using the ASW method [9, 10]. Kulatov also carried out band calculation using the LMTO method. Gross features of density of states obtained by them are similar to the result shown in Fig. 5.5.

5.2.3 Ferromagnetic State of MnAlGe and MnGaGe

In these compounds, atomic planes of Mn at M(I)-site are separated by two planes consisting of Al (or Ga) and Ge, which suggests a two-dimensional character in magnetic properties. Observed magnetic moment is much smaller than moment expected for localized Mn orbitals. Therefore, we can regard to some extent these compounds as an itinerant electron ferromagnets with two-dimensional character.

Figure 5.6 shows calculated density of states for MnAlGe [11]. MnAlGe has density of states similar to that of MnGaGe. Hybridization between *d*-orbitals of Mn and *p*-orbitals of Al (Ga) and Ge is small. As shown in Table 5.3, calculated moment of MnAlGe is in good agreement with the experimental value, while the agreement is not good for MnGaGe. According to the band calculations, very small magnetic moment is induced at Al (Ga) and Ge atomic site in the opposite direction of the Mn-moment.

We found, as shown in Fig. 5.7, three Fermi surfaces of the down-spin bands consisting of two electron surfaces and a cylindrical hole surface around the *Z*-axis, and four Fermi surfaces of the up-spin bands consisting of a small hole pocket around the R-point, a big hole pocket, and two electron pockets near the M-point. The Fermi

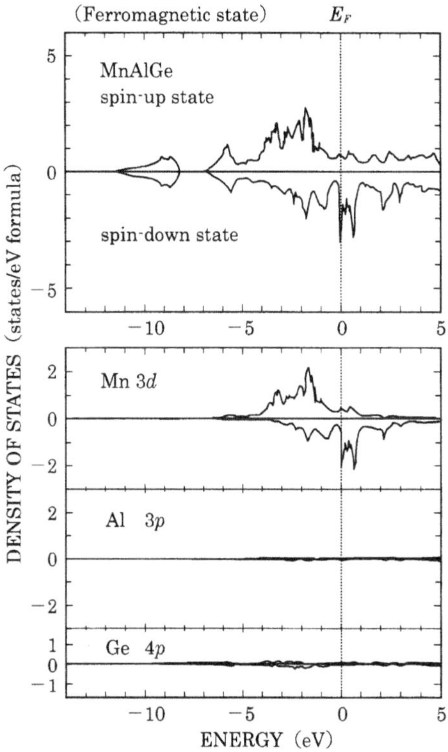

Fig. 5.6. Density of states of ferromagnetic bands of MnAlGe

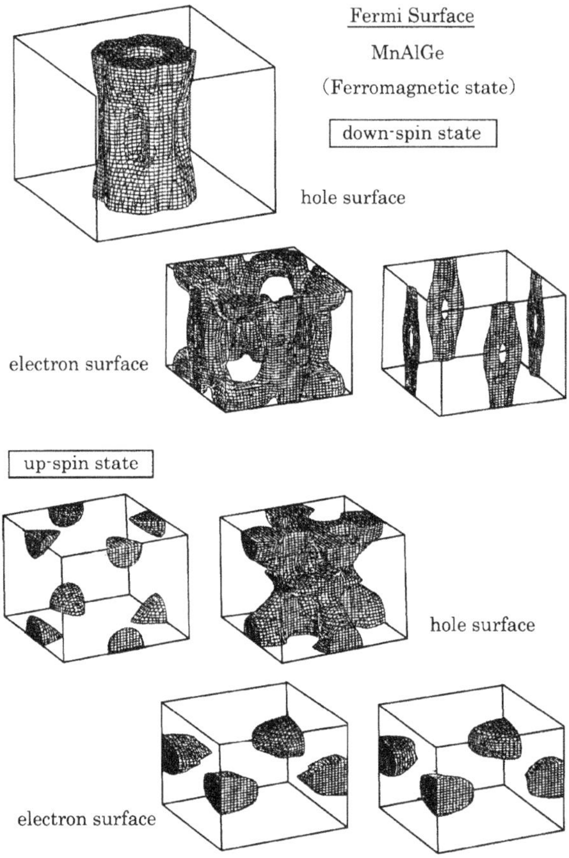

Fig. 5.7. Fermi surfaces of ferromagnetic bands of MnAlGe

surfaces of the down-spin bands has large contribution to $\rho(E_F)$. We expect that the strong two-dimensional character of the Fermi surfaces of the down-spin bands will be observed as anisotropy of transport properties such as resistivity.

5.2.4 Antiferromagnetic Bands of Cr_2As, Mn_2As, and Fe_2As

We carried out band calculations for antiferromagnetic state of these compounds. As seen in Fig. 2.2 of Part I and Fig. 5.1 of Part II, magnetic moments of Cr(1) and Cr(2) of Cr_2As are in the c-plane and antiparallel, and so Cr(1) and Cr(2) are regarded as different sites. On the other hand, in Mn_2As and Fe_2As, the sites 1 and 2 are equivalent because moment at these sites are parallel. Magnetic moments at the M(I)-site and adjacent M(II)-site are antiparallel in Mn_2As and parallel in Fe_2As. The spin arrangements for the compounds with Cu_2Sb-type structure are illustrated in Fig. 2.2 of Part I.

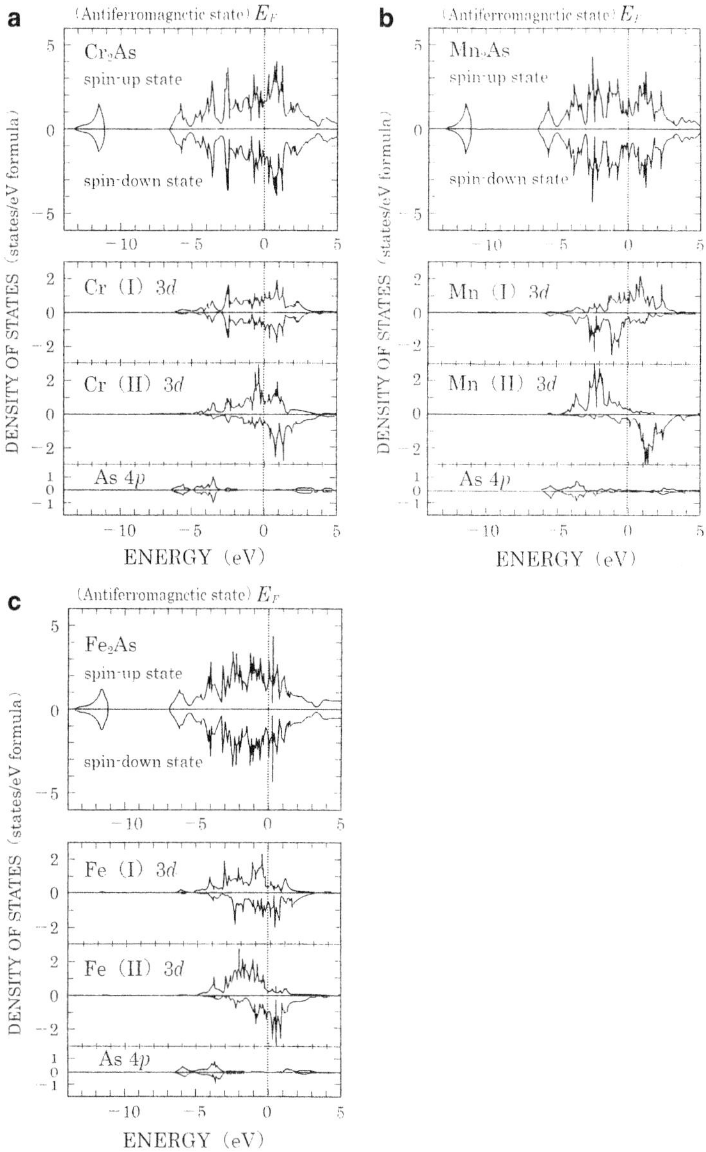

Fig. 5.8. Total and local densities of states for (**a**) Cr_2As, (**b**) Mn_2As, and (**c**) Fe_2As in antiferromagnetic state

Figure 5.8 shows total and local densities of states for MAs with M = Cr, Mn, and Fe. It is noteworthy that in Cr_2As the splitting of Cr(II)-bands is about 1 eV, whereas there is little splitting of Cr(I)-bands. The very small splitting in Cr(I)-local density of states results in the very small magnetic moment of Cr(I), which is related to the antiferromagnetic order in the c-plane. As for Fe_2As and Mn_2As, mixing between

3*d*-states at the M(I)- and the M(II)-sites in Mn_2As is smaller than that in Fe_2As. Calculated moments are compared with the experimental data in Table 5.3. As shown in Table 5.3, magnetic moment of M(II) is larger than that of M(I) in the respective compounds. Agreement between calculation and experiment is good for Cr_2As and Fe_2As but not so good for Mn_2As.

5.2.5 Magnetic Ordering of Cu_2Sb-Type Compounds

The Cu_2Sb-type compounds except for MnAlGe and MnGaGe have four magnetic atoms in a unit cell, and therefore various types of magnetic structures have been observed. We have studied such various magnetic structures from the viewpoint of itinerant electrons. For this picture we investigated instability of paramagnetic phase within the theoretical framework described in Sect. 4.6. In this section we explain only the calculated results.

Figure 5.9 shows the magnetic phase diagram calculated for $Q = (0,0,0)$ and $Q = (0,0,\pi/c)$, respectively. The dashed-curves express the relation between the critical values of intra-atomic Coulomb interaction divided by a transfer integral,

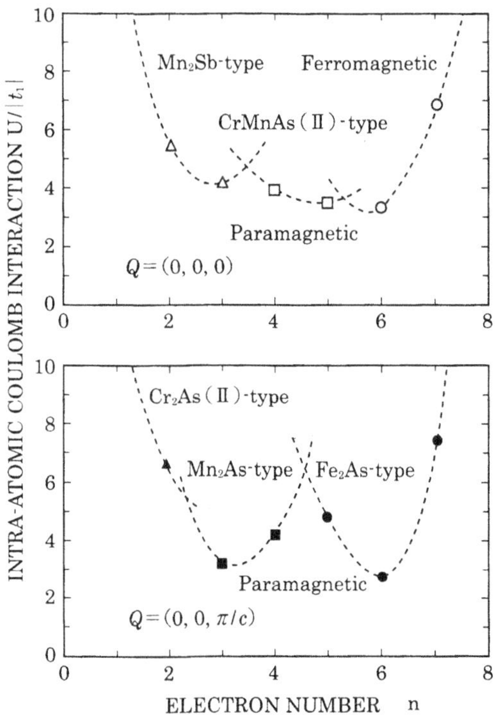

Fig. 5.9. A magnetic phase diagram of the Cu_2Sb-type compounds. The symbols *open circle*, *open triangle*, *open square*, etc. denote critical values of intra-atomic Coulomb interaction calculated for each magnetic ordering (see also the text). The magnetic orders expressed by CrMnAs(II)-type and Cr_2As-type are hypothetical ones for the theoretical calculation [14]

$U_c/|t_1|$. $U_c/|t_1|$ is plotted as a function of average electron number n. As we have noted in Sect. 4.6, if U is larger than U_c paramagnetic phase is unstable. The magnetic structure that corresponds to the smallest value of U_c is most likely to realize.

From the calculations, it is indicated that Mn_2Sb-type, CrMnAs(II)-type, and ferromagnetic ordering are theoretically possible for $Q = (0,0,0)$. For $Q = (0,0,\pi/c)$, the Cr_2As(II)-type, Mn_2As-type, and Fe_2As-type can theoretically realize. The magnetic orderings actually observed are included in Fig. 5.9.

5.2.6 Photoemission and Inverse Photoemission

It is possible to obtain a rough estimation of density of states below and above the Fermi level by photoemission and inverse photoemission measurements, respectively. Suga and Kimura measured photoemission and inverse photoemission spectra of Cu_2As, Mn_2As, Fe_2As, MnAlGe, and MnGaGe. Their experimental data are shown in Fig. 5.10 [12, 13] together with the calculated density of states summed over the up-spin and down-spin bands [14]. Overall features of the spectra correspond well to the calculations. For Mn_2Sb, the spectrum has remarkable peaks at about 3 eV below the Fermi level and 2 eV above the Fermi level, while in MnAlGe

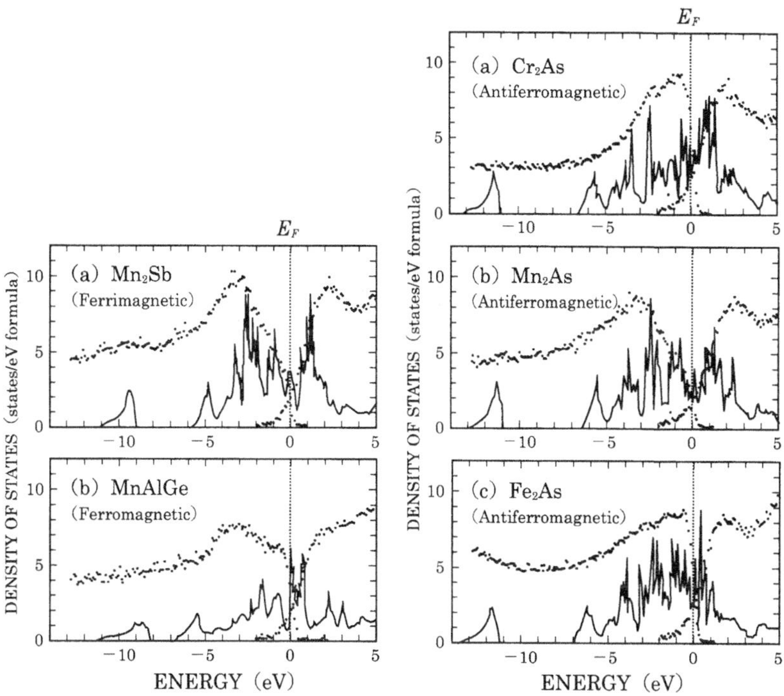

Fig. 5.10. Densities of states determined by photoemission and inverse photoemission spectra (*dots*) and calculated densities of states (*solid lines*) for the Cu_2Sb-type compounds

there is no clear peak but broad hump with the width about 4.5 eV lies below Fermi level. Such difference between the spectra of Mn_2Sb and MnAlGe can be attributed to the absence of Mn at M(II) site in the case of MnAlGe.

From the spectra of Cr_2As, Mn_2As, and Fe_2As, we see that energy difference between the peaks above and below the Fermi level in Cr_2As and Fe_2As is smaller than the energy difference in Mn_2As.

References

[1] K. Adachi, S. Ogawa, *Landolt-Börnstein New Series III/***27a**, ed. by H.P.J. Wijn (Springer, Berlin, 1988), p. 265

[2] M. Shirai, K. Motizuki, *Recent Advances in Magnetism of Transition Metal Compounds*, ed. by A. Kotani, N. Suzuki (World Scientific, Singapore, 1993), p. 67

[3] T. Takeda, J. Kübler, J. Phys. F Metal Phys. **9**, 661 (1979)

[4] O. Gunnarsson, B.I. Lundqvist, Phys. Rev. B **13**, 4274 (1976)

[5] E. Kulatov, L. Vinokurova, K. Motizuki, *Recent Advances in Magnetism of Transition Metal Compounds*, ed. by A. Kotani, N. Suzuki (World Scientific, Singapore, 1993), p. 56

[6] T. Chônan, A. Yamada, K. Motizuki, J. Phys. Soc. Jpn. **60**, 1638 (1991)

[7] T. Ito, M. Shirai, K. Motizuki, J. Phys. Soc. Jpn. **61**, 2202 (1992)

[8] M. Suzuki, M. Shirai, K. Motizuki, J. Phys. Condens. Matter **4**, L33 (1992)

[9] J.H. Wijngaard, C. Haas, R.A. de Groot, Phys. Rev. B **45**, 5395 (1992)

[10] C. Haas, R.A. de Groot, *Recent Advances in Magnetism of Transition Metal Compounds*, ed. by A. Kotani, N. Suzuki (World Scientific, Singapore, 1993), p. 78

[11] K. Motizuki, T. Korenari, M. Shirai, J. Magn. Magn. Mater. **104–107**, 1923 (1992)

[12] A. Kimura, S. Suga, H. Matsubara, T. Matsushima, Y. Saito, H. Daimon, T. Kaneko, T. Kanomata, Solid State Commun. **81**, 707 (1992)

[13] S. Suga, A. Kimura, *Recent Advances in Magnetism of Transition Metal Compounds*, ed. by A. Kotani, N. Suzuki (World Scientific, Singapore, 1993), p. 91

[14] M. Suzuki, M. Shirai, K. Motizuki, J. Phys. Condens. Matter **4**, L33 (1992)

Index

GPSR Compliance

The European Union's (EU) General Product Safety Regulation (GPSR) is a set of rules that requires consumer products to be safe and our obligations to ensure this.

If you have any concerns about our products, you can contact us on ProductSafety@springernature.com

In case Publisher is established outside the EU, the EU authorized representative is:

Springer Nature Customer Service Center GmbH
Europaplatz 3
69115 Heidelberg, Germany

Batch number: 10584666

Printed by Printforce, the Netherlands